I0106150

Willy's GPW
1/4 Ton Truck
Manuals

TM 9-1803A

Engine and Engine Accessories Maintenance Manual

TM 9-1803B

Power Train, Body and Frame Maintenance Manual

Use the black index marks to quickly find each manual in this book.

edited by
Brian Greul

The GPW Model MB, commonly referred to as the Jeep is probably the most ubiquitous American military vehicle ever produced. Simple, rugged, and capable of hard work it began as a World War II vehicle and descendant vehicles are still produced today as passenger vehicles.

This book is intended to support enthusiasts and their restoration efforts by providing a professionally printed, 8.5x11 compilation of the key manuals for this vehicle.

Every effort has been made to faithfully reproduce the document while cleaning up the pages to make them usable to you the reader. However, we are dealing with original works that have been electronically preserved from nearly 80 years ago. There are a number of artifacts in the source documents. Your understanding is appreciated.

An 8.5x11 3 hole punched loose leaf copy may be purchased for your 3 ring binder. Email books@ocotillopress.com for current information.

Should you have suggestions or feedback on ways to improve this book please send email to Books@OcotilloPress.com

---

Edited 2021 Ocotillo Press
ISBN 978-1-954285-12-5

Printed in the United States of America

Ocotillo Press
Houston, TX 77017
Books@OcotilloPress.com

WAR DEPARTMENT TECHNICAL MANUAL

TM 9-1803A

ORDNANCE MAINTENANCE

Engine and Engine Accessories For ¼-Ton 4x4 Truck

(Willys-Overland Model MB and Ford Model GPW)

WAR DEPARTMENT *24 FEBRUARY 1944*

CONTENTS

ORDNANCE MAINTENANCE — ENGINE AND ENGINE ACCESSORIES FOR ¼-TON 4x4 TRUCK (WILLYS-OVERLAND MODEL MB AND FORD MODEL GPW)

CHAPTER 1

INTRODUCTION

1. SCOPE.

a. The instructions contained in this manual are for the information and guidance of personnel charged with the maintenance and repair of the 4-cylinder engine used in the Willys MB and Ford GPW ¼-ton 4 x 4 Trucks. These instructions are supplementary to field and technical manuals prepared for the using arms. This manual does not contain information which is intended primarily for the using arms, since such information is available to ordnance maintenance personnel in 100-series TM's or FM's.

b. This manual contains a description of, and procedure for inspection, removal, disassembly, repair, and rebuilding of the engine.

c. TM 9-803 contains information and guidance for the using arms and first and second echelons.

d. TM 9-1803B contains information for removal, inspection, repair, rebuild, assembly, and installation of the power train and chassis.

e. TM 9-1825B contains information for the maintenance of the Auto-Lite electrical equipment used on this vehicle.

f. TM 9-1826A contains information for the maintenance of the Carter carburetor used on this vehicle.

g. TM 9-1827C contains information for the maintenance of the Wagner hydraulic brake system used on this vehicle.

h. TM 9-1828A contains information for the maintenance of the A. C. fuel pump used on this vehicle.

i. TM 9-1829A contains information for the maintenance of the speedometer used on this vehicle.

j. This manual includes engine ordnance maintenance instructions from the following Quartermaster Corps 10-series technical manuals. Together with TM 9-803 and TM 9-1803B, this manual supersedes them:

(1) TM 10-1103, 20 August 1941.

(2) TM 10-1207, 20 August 1941.

(3) TM 10-1349, 3 January 1942.

(4) TM 10-1513, Change 1, 15 January 1943.

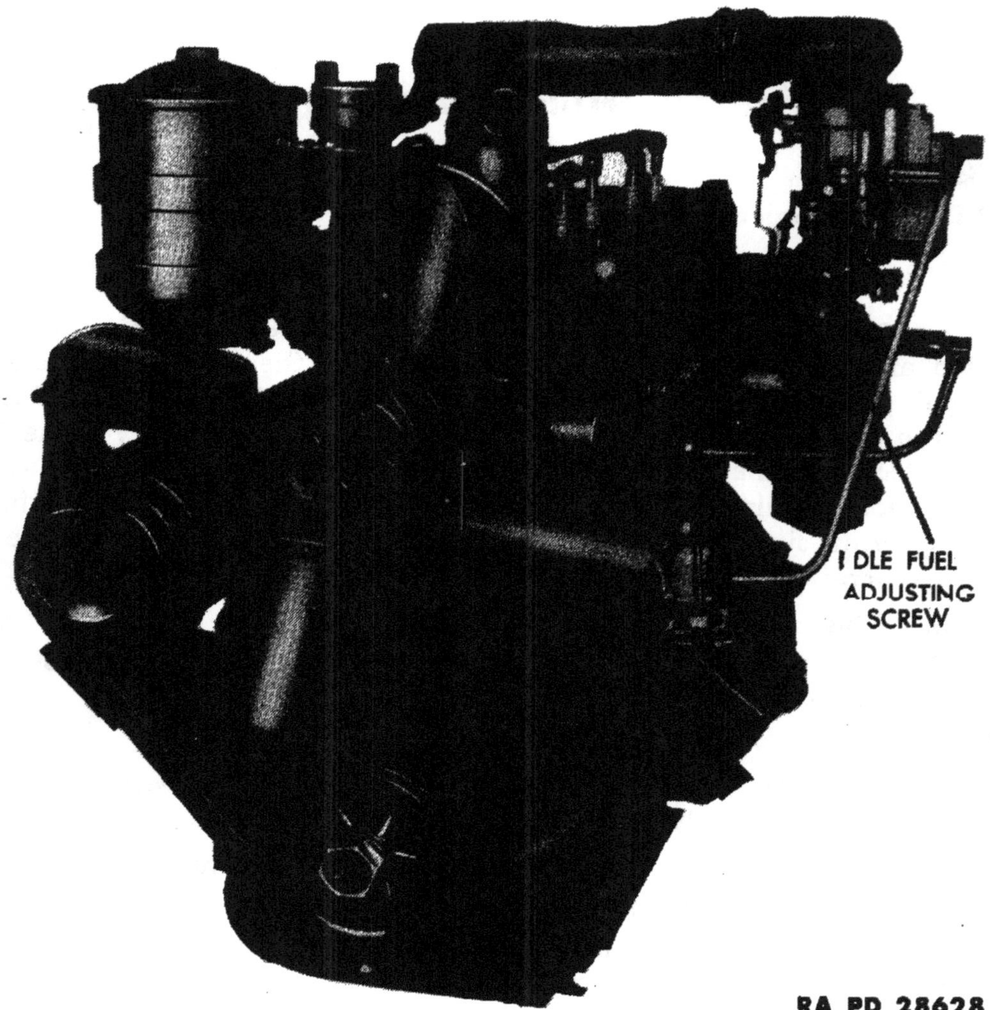

Figure 1 — Front View of Engine

2. MWO AND MAJOR UNIT ASSEMBLY REPLACEMENT RECORD.

a. **Description.** Every vehicle is supplied with a copy of AGO Form No. 478 which provides a means of keeping a record of each MWO (FSMWO) completed or major unit assembly replaced. This form includes spaces for the vehicle name and U.S.A. registration number, instructions for use, and information pertinent to the work accomplished. It is very important that the form be used as directed, and that it remain with the vehicle until the vehicle is removed from service.

RA PD 28665

Figure 2 — Left Side View of Engine

RA PD 28664

Figure 3 — Right Side View of Engine

INTRODUCTION

description of the work completed, and must initial the form in the columns provided. When each modification is completed, record the date, hours and/or mileage, and MWO number. When major unit assemblies, such as engines, transmissions, transfer cases, are replaced, record the date, hours and/or mileage, and nomenclature of the unit assembly. Minor repairs and minor parts and accessory replacements need not be recorded.

c. **Early Modifications.** Upon receipt by a third or fourth echelon repair facility of a vehicle for modification or repair, maintenance personnel will record the MWO numbers of modifications applied prior to the date of AGO Form No. 478.

ORDNANCE MAINTENANCE — ENGINE AND ENGINE ACCESSORIES FOR ¼-TON 4x4 TRUCK (WILLYS-OVERLAND MODEL MB AND FORD MODEL GPW)

CHAPTER 2

ENGINE

Section I

DESCRIPTION AND DATA

3. DESCRIPTION.

a. The engine used in the ¼-ton 4 x 4 Truck is the 4-cylinder, L-head, gasoline-type (figs. 1, 2, and 3), equipped with a counter-balanced crankshaft. The camshaft is operated off the crankshaft through a timing chain (fig. 40). The oil pump and distributor operate off the camshaft.

4. DATA.

Type	L-head
Numbers of cylinders	4
Bore and stroke	3.125 x 4.375 in.
Piston displacement	134.2 cu in.
Compression ratio	6.48 to 1
Max. brake horsepower	54 at 4,000
Compression (lb per sq in. at 185 rpm)	111
SAE horsepower	15.63
Maximum torque	105 ft-lb at 2,000 rpm
Firing order	1-3-4-2

Section II

ENGINE REMOVAL FROM VEHICLE

5. REMOVAL FROM VEHICLE.

a. General. Unhook the two hood clamps, raise the hood, and lay it against the windshield. Drain the coolant from the radiator and the engine by opening the radiator drain cock and the drain cock

RADIATOR
BOND STRAP
RADIATOR BOLT
RADIATOR DRAIN COCK
OIL PAN
OIL PAN DRAIN PLUG
EXHAUST PIPE
STAY CABLE
RA PD 28748
BOND STRAP
RADIATOR BOLT

Figure 4 — Underside View of Engine Installed in Vehicle

ORDNANCE MAINTENANCE — ENGINE AND ENGINE ACCESSORIES FOR ¼-TON 4x4 TRUCK (WILLYS-OVERLAND MODEL MB AND FORD MODEL GPW)

Figure 5 — Right Side View of Engine Installed in Vehicle

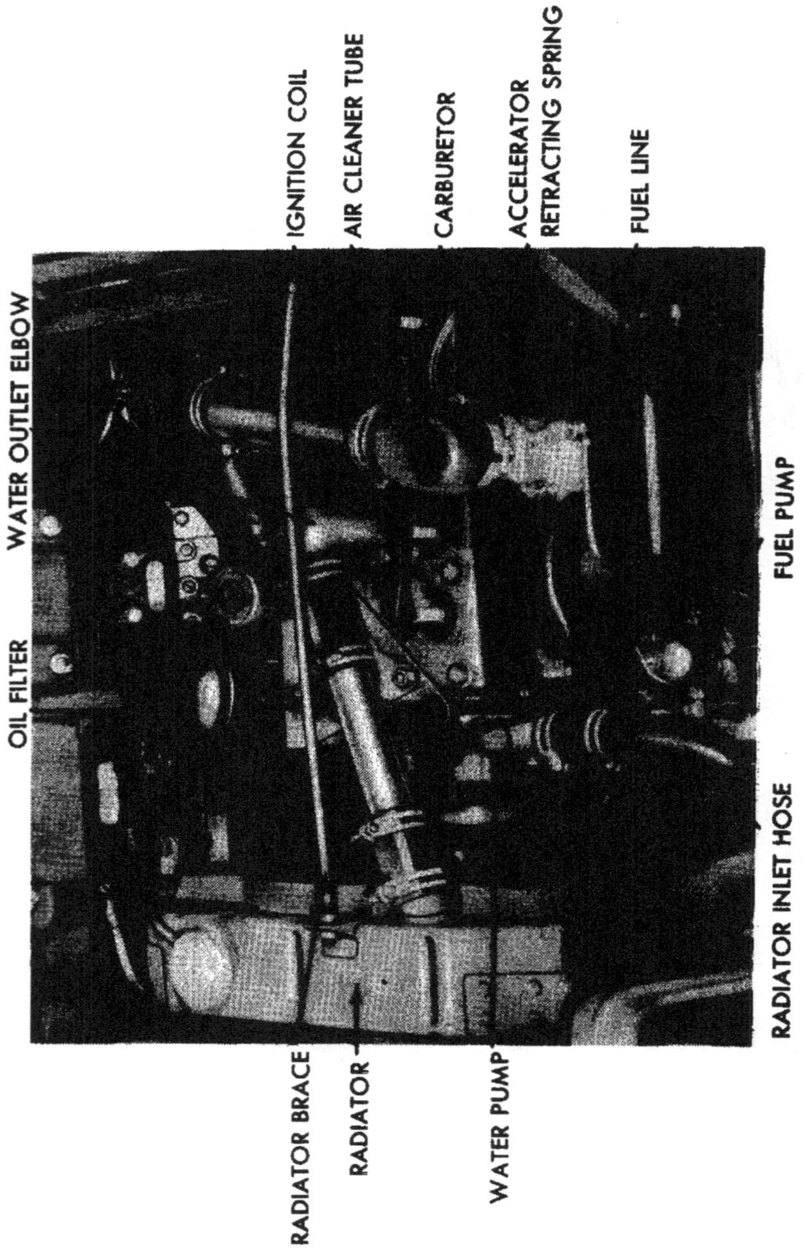

RA PD 28814

Figure 6 — Side View of Engine Compartment

located on the right-hand side of the engine. Remove the oil pan drain plug and drain the engine oil. Some variation exists in the location of the various bond straps used to eliminate radio interference on these vehicles. Disregard references to bond straps in the following instructions if they are not present on the particular vehicle being worked on. If bond straps are found in locations other than those mentioned in the following procedure, they should be disconnected, if they prevent removal of the engine.

b. **Remove Battery.** Loosen the two battery cable bolts, and disconnect both cables. Loosen the battery brace wing nut on the fender. Remove the two battery hold-down frame wing nuts (fig. 5). Move the battery brace to one side, and remove the battery hold-down frame. Lift the battery from the vehicle.

c. **Remove Radiator.** Remove the nut and lock washer from the front and rear of the radiator brace, and remove the brace. Loosen the two front outlet radiator hose clamps, and slide the hose back on the metal tubing. Loosen the rear radiator outlet hose clamp, and remove the hose. Loosen the radiator hose clamps on the inlet hose at the water pump, also the one on the radiator and remove the radiator inlet hose. Working from underneath the vehicle, remove the two nuts, flat washers, and bond straps from the radiator bolts (fig. 4). Remove the two nuts and flat washers from the radiator bolts. Lift the radiator from the vehicle and remove the two radiator pads.

d. **Disconnect Oil and Water Temperature Gages.** Disconnect the oil gage line at the flexible oil line, located at the left-hand side of the engine. Disconnect the water temperature gage (engine unit) at the right-hand side of the cylinder head (fig. 5).

e. **Remove Air Cleaner Hose** (fig. 6). Loosen the hose clamps on the carburetor air cleaner and oil filler pipe, and remove the air cleaner hose.

f. **Disconnect Electrical Wires and Bond Straps.** Disconnect the field, armature, and ground wires at the generator. Disconnect the primary wire running from the dash to the coil, at the coil. Disconnect the bond strap at the rear of the cylinder head. Disconnect the ground strap at each front engine support. Disconnect the cranking motor cable at the cranking motor.

g. **Remove Cranking Motor.** Remove the cap screw that holds the cranking motor bracket to the cylinder block. Remove the two cap screws that hold the cranking motor to the clutch housing, and slip the cranking motor from the engine.

Figure 7 — Lifting Engine from Vehicle

h. Disconnect Choke and Throttle Controls. Remove the nut and bolt on the choke and throttle hold-down bracket. Loosen the set screw on the carburetor choke lever, and remove the choke control cable. Loosen the set screw on the throttle control cable and remove the throttle control cable. Disconnect the throttle control at the accelerator pedal in the driver's compartment.

i. Disconnect Exhaust Pipe. Remove the nut, bolt, and cap screw that hold the exhaust pipe to the exhaust manifold. Pry the exhaust pipe from the exhaust manifold.

j. Disconnect Front Engine Supports. Remove the two nuts and bolts from each front engine support.

k. Remove Stay Cable and Clutch Housing Bolts. Remove the two engine stay cable nuts at the front crossmember, and remove the stay cable (fig. 4). Remove the 10 cap screws and bolts from the clutch housing.

l. Remove Engine From Vehicle. Install a suitable lifting sling or rope on the engine (fig. 7). Raise the engine high enough to release the weight on the front engine supports. Pull the engine forward until it is free from the clutch housing, and lift the engine from the vehicle (fig. 7).

Section III

DISASSEMBLY OF ENGINE INTO SUBASSEMBLIES

6. PRELIMINARY OPERATIONS.

a. General. If the clutch housing was removed with the engine, start the procedure beginning with subparagraph b below. If the clutch housing was not removed with the engine, remove the cranking motor (par. 5 **g**), remove the rest of the clutch housing bolts or cap screws, and remove the clutch housing from the engine.

b. Remove Carburetor (fig. 6). Remove the fuel line connecting the carburetor and fuel pump. Remove the accelerator return spring from the careburetor and accelerator lever. Remove the two carburetor hold-down nuts, lock washers, and accelerator return spring clip.

c. Remove Fuel Pump (fig. 6). Disconnect the other fuel line at the fuel pump. Remove the two cap screws and lock washers that hold the fuel pump to the cylinder block, and remove the fuel pump.

d. Remove Distributor (fig. 5). Pull the spark wires off the spark plugs, and slide the wires out of the air filter tube bracket. Remove the primary and secondary wires from the ignition coil. Remove the distributor hold-down screw, and lift the distributor and wires from the cylinder block.

e. Remove Oil Filter (fig. 5). Disconnect the inlet oil line on the left-hand side of the cylinder block, and the outlet oil line on the engine front cover. Remove the cap screw that holds the oil filler pipe to the oil filter bracket. Remove the three cylinder head nuts

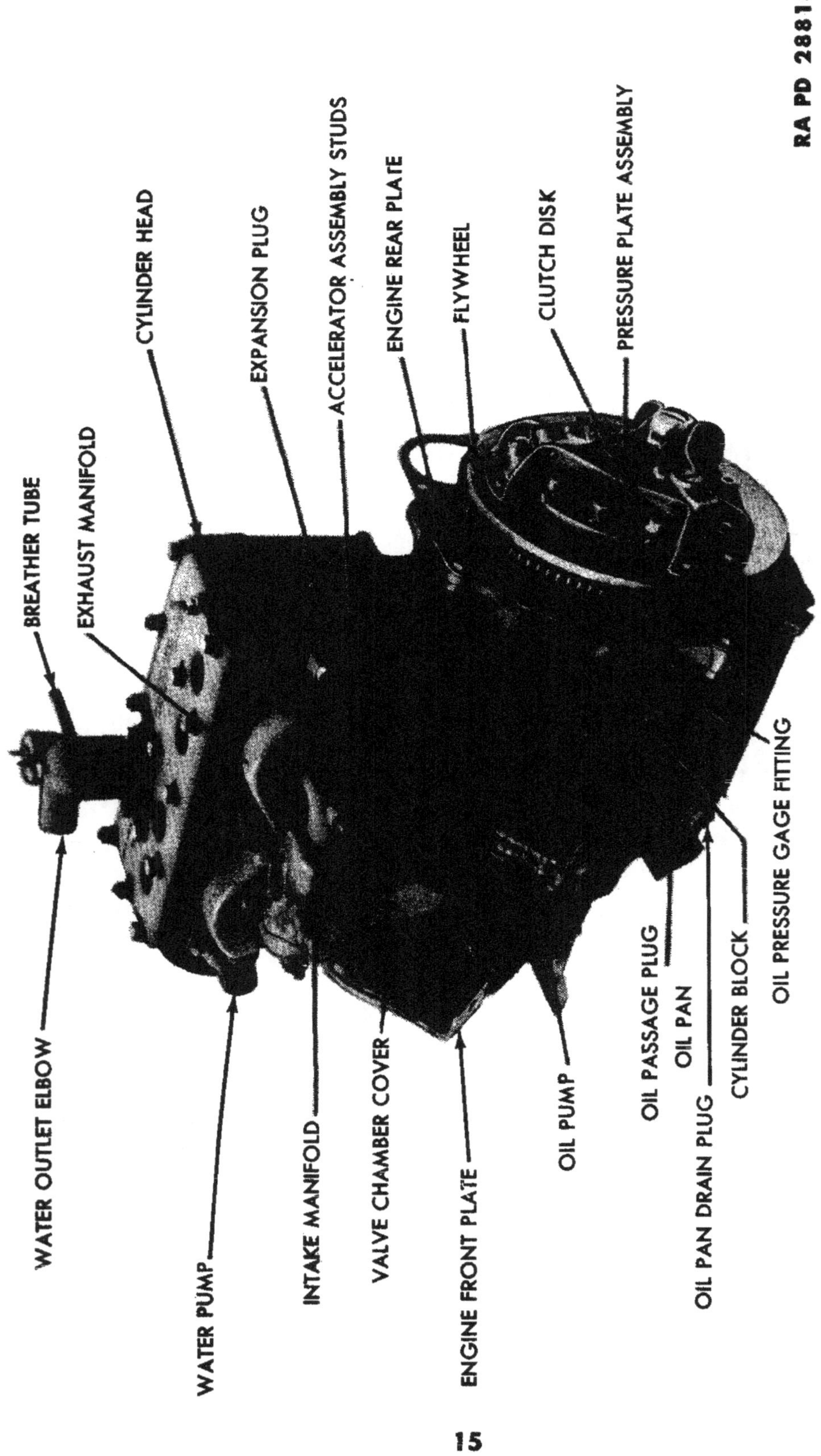

Figure 8 — Three-quarter Left Rear View of Stripped Engine

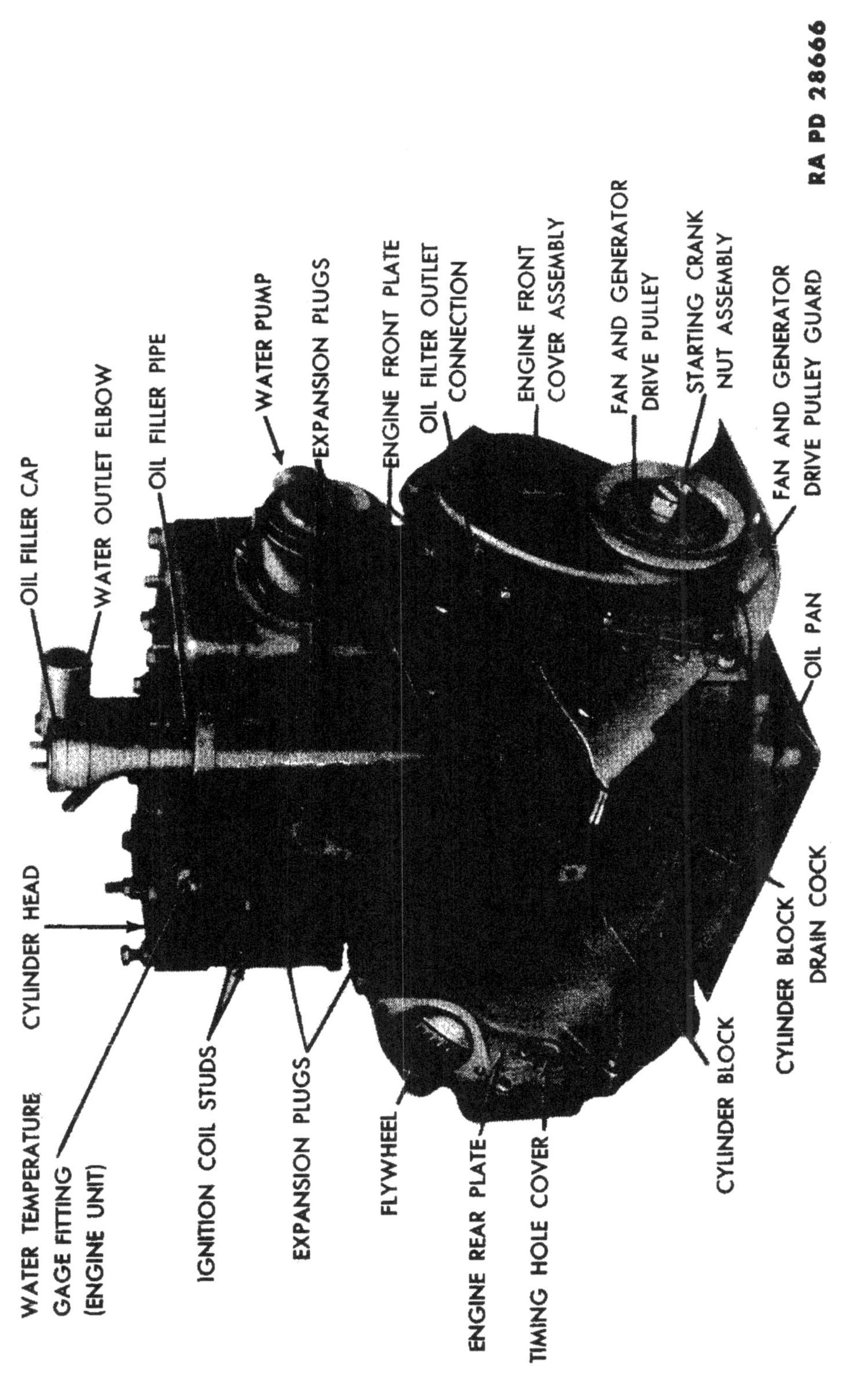

RA PD 28666

Figure 9 — Three-quarter Right Front View of Stripped Engine

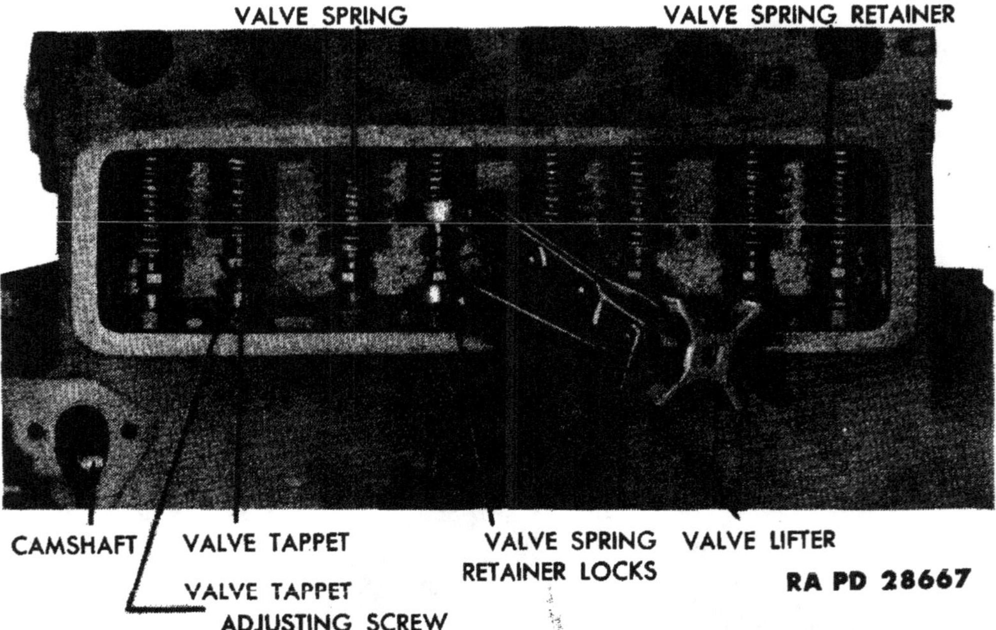

Figure 10 — Removing Valve Spring Retainer Locks, Using Valve Lifter (41-L-1410)

that hold the oil filter bracket to the cylinder head, and remove the oil filter.

f. Remove Generator and Generator Support Bracket. Pull up on the generator adjusting bracket, raise the generator to release the tension on the fan belt, and remove the belt. Remove the two bolts that hold the generator to the support bracket, and remove the generator. Remove the two cap screws that hold the generator support bracket to the cylinder block, and remove the generator support bracket.

g. Remove Ignition Coil. Remove the two nuts and lock washers that hold the ignition coil to the cylinder block, and remove the ignition coil and bond strap.

h. Remove Fan. Remove the four cap screws and lock washers that hold the fan to the water pump, and remove the fan.

7. DISASSEMBLY OF STRIPPED ENGINE.

a. Remove Water Pump (fig. 9). Remove the four cap screws and lock washers that hold the water pump to the cylinder block, and remove the water pump.

b. Remove Intake and Exhaust Manifold (fig. 8). Remove the ventilating tube that connects the intake manifold and valve chamber cover. Remove the seven nuts, and lift the intake and exhaust manifold off the engine.

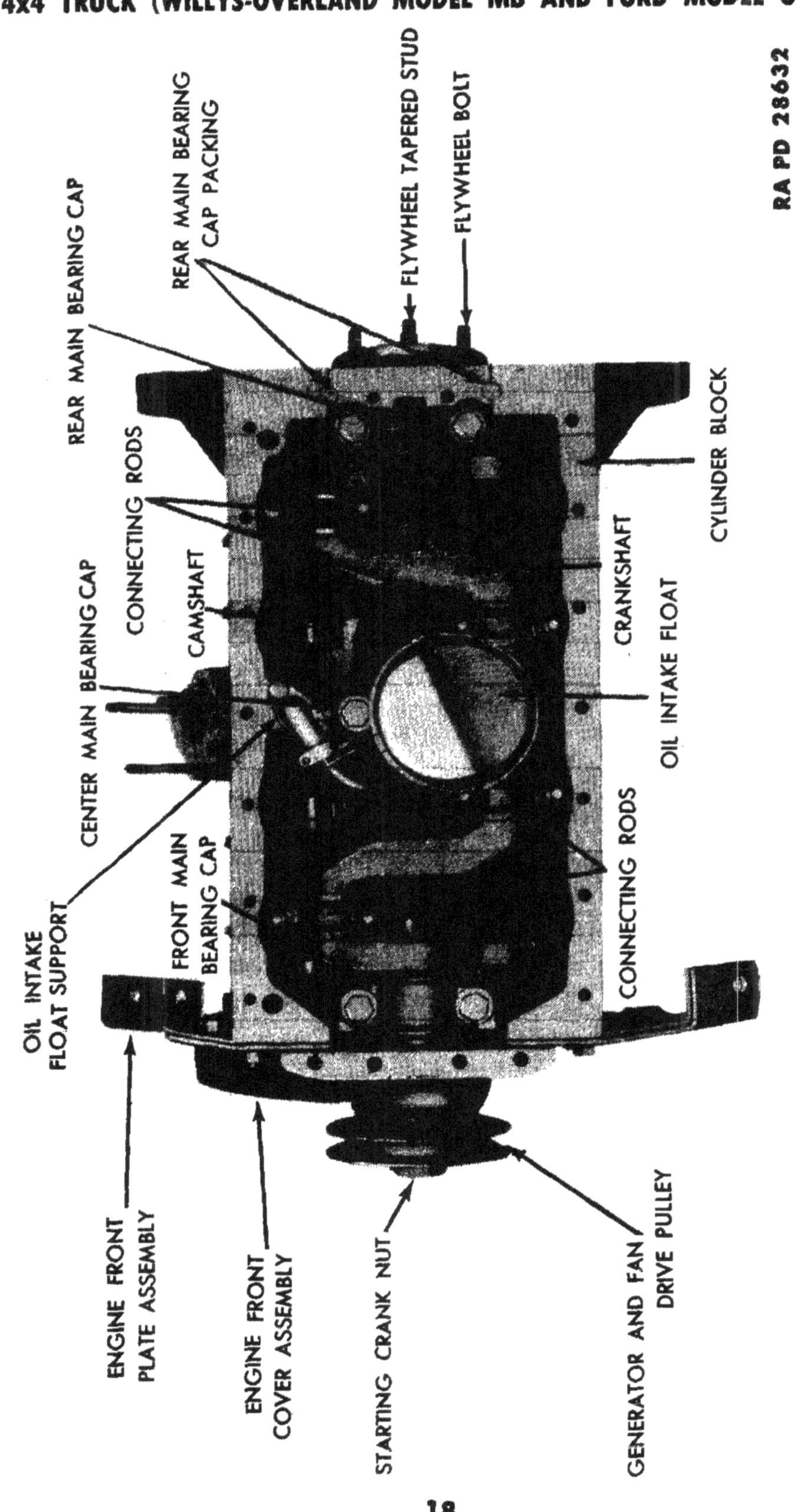

Figure 11 — Underside View of Engine with Oil Pan Removed

c. **Remove Water Outlet Elbow** (fig. 8). Remove the three nuts that hold the water outlet elbow to the cylinder head, and remove the water outlet elbow and thermostat. Remove the thermostat retainer and thermostat.

d. **Remove Clutch Disk** (fig. 8). Loosen the six pressure plate bracket cap screws in sequence, a little at a time, to prevent distortion of the pressure plate bracket. Remove the six cap screws, pressure plate, and clutch disk.

e. **Remove Flywheel** (fig. 8). Remove the six nuts and lock washers that hold the flywheel to the crankshaft. Tap the flywheel off the crankshaft with a brass hammer. Lift the rear engine plate from the engine.

f. **Remove Cylinder Head** (fig. 9). Remove the remaining cap screws that secure the head to the cylinder block, and remove the cylinder head.

g. **Remove Valves and Springs** (fig. 10). Remove the two cap screws and crankcase ventilator assembly from the valve chamber cover, and remove the cover. With a valve lifter (41-L-1410) inserted between the valve tappet and valve spring retainer, raise the valve springs that are in closed position, and remove the valve spring retainer locks (fig. 10). Turn the crankshaft until those valves which are open become closed, and remove the rest of the valve spring retainer locks. Remove the valves and place them in a valve carrying board, so that they can be identified as to cylinders from which they were removed. Compress the valve spring with the valve lifter on each valve tappet that is in the closed position, and pull the spring off the valve guide. Turn the crankshaft until the tappets are in a closed position, and remove the rest of the valve springs.

h. **Remove Oil Pan and Oil Intake Float.** Turn the engine on its side, and remove the cap screws that secure the oil pan and fan pulley guard to the cylinder block. Remove the fan pulley guard and oil pan. Remove the two cap screws from the oil intake float (fig. 11), and remove the oil intake float.

i. **Remove Camshaft Sprocket and Camshaft.** Remove the eight nuts and bolts that secure the engine front cover and engine front plate to the cylinder block and remove the cover. Remove the camshaft thrust plunger and spring. Straighten the tabs on the four camshaft sprocket cap screw lock washers (fig. 25), and remove the four cap screws and lock washers. Lift the camshaft sprocket and the camshaft drive link chain off the camshaft. Remove the camshaft thrust washer. Lay the cylinder block on its side. Pull all the valve tappets toward the top of the cylinder block. Pull the cam-

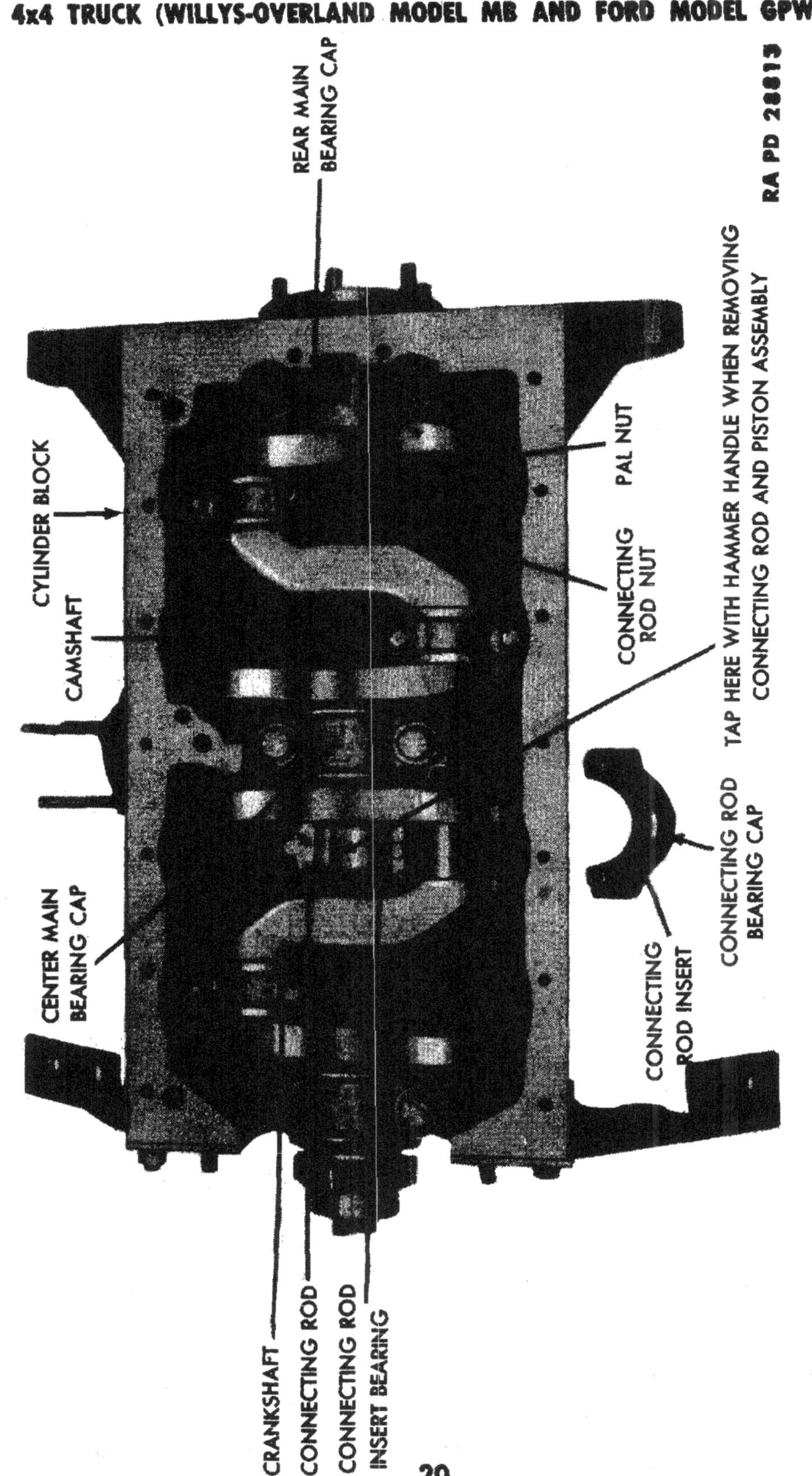

Figure 12 — Connecting Rod and Piston Assembly Removal

shaft out of the cylinder block, and remove the valve tappets. Remove the three cap screws that hold the engine front plate to the cylinder block, and remove the plate.

j. **Remove Piston and Connecting Rod Assemblies** (fig. 12). Remove the two pal nuts, connecting rod nuts, and connecting rod bearing cap from each connecting rod. Remove all carbon from the top of the cylinder walls. Tap the connecting rod and piston assembly out of the cylinder block with the handle end of a hammer (fig. 12). Install the connecting rod bearing caps on the rods in same position as originally installed, to prevent later improper mating of parts.

k. **Remove Crankshaft.** Remove the two cap screws from each main bearing cap (fig. 11), and remove the three main bearing caps. Lift the crankshaft from the cylinder block.

Section IV

DISASSEMBLY, CLEANING, INSPECTION, REPAIR, AND ASSEMBLY OF SUBASSEMBLIES

8. CYLINDER BLOCK, HEAD, AND OIL PAN.

a. **Cleaning.** Strip off all old gaskets and sealing compound from all machined surfaces. Remove plugs, and clean all oil passages in the cylinder block with steam or compressed air. Scrape the carbon from the cylinder block and head. Clean the cylinder block, head, and oil pan thoroughly with dry-cleaning solvent.

b. **Inspection and Repair.**

(1) OIL PAN (fig. 13). An oil pan with stripped threads in the drain plug opening, or an oil pan that is badly dented or deformed, must be replaced.

ORDNANCE MAINTENANCE — ENGINE AND ENGINE ACCESSORIES FOR 1/4-TON 4x4 TRUCK (WILLYS-OVERLAND MODEL MB AND FORD MODEL GPW)

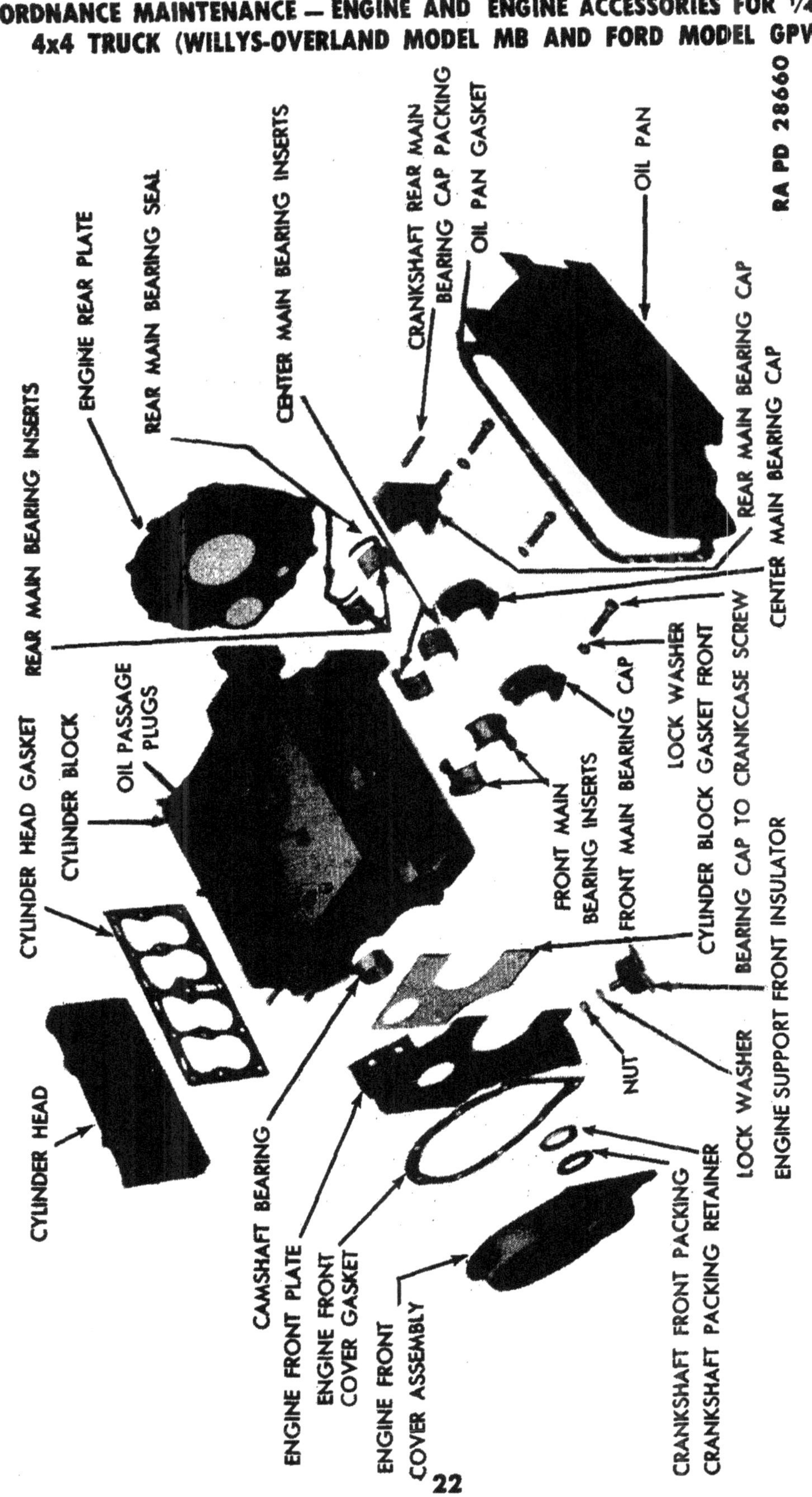

Figure 13 — Cylinder Block, Head, Oil Pan, and Bearings, Disassembled

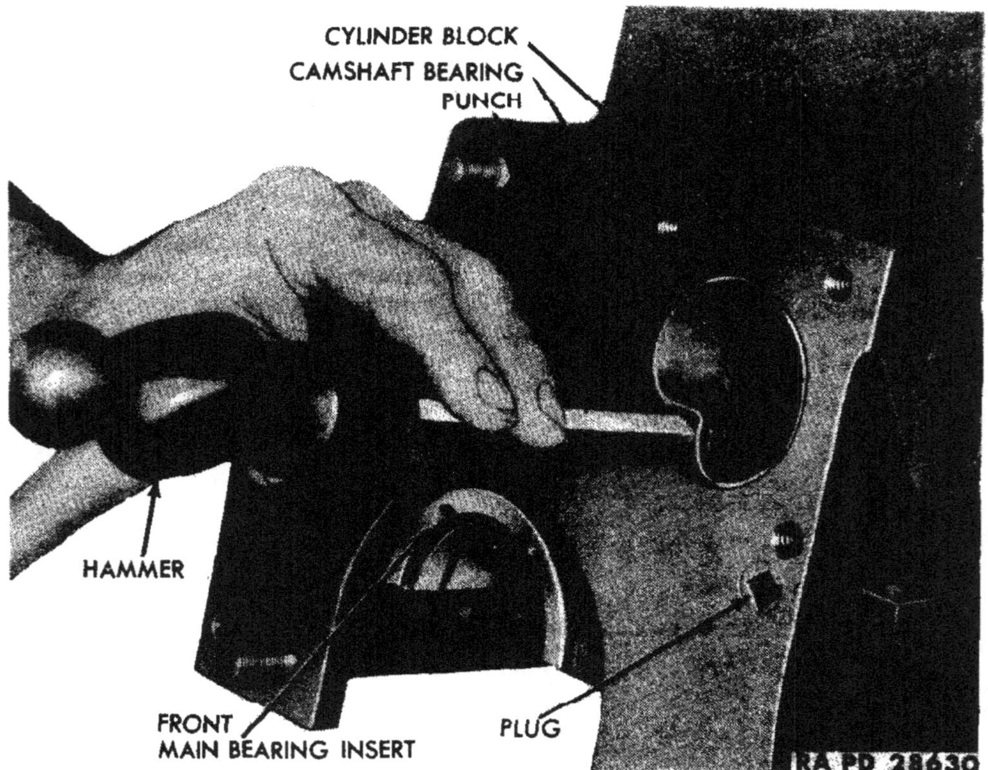

Figure 14 — Driving Camshaft Bearing from Cylinder Block

(2) CYLINDER HEAD (fig. 13). A cracked or warped cylinder head, or a cylinder head with stripped threads in the spark plug holes, must be replaced.

(3) CYLINDER BLOCK (fig. 13). A cracked or damaged cylinder block must be replaced. All loose expansion plugs (fig. 9) or damaged studs must be replaced (step (4) below). A scored, ridged, discolored, or excessively worn, front camshaft bearing (fig. 13) (worn to more than 2.190 in. inside diameter) must be replaced (step (5) below). Measure the other three camshaft bearings with a micrometer caliper. If the bearings are larger than 2.128 inches for the front intermediate, 2.1395 inches for the rear intermediate, or 1.628 inches for the rear bearing, the cylinder block must be replaced. Measure the cylinder bores with a micrometer caliper and telescope gage. If any of the cylinders has a taper of more than 0.010 inch, or an out-of-round condition of more than 0.005 inch, the cylinders must be rebored to 0.020 or 0.030 inch oversize. If cylinder walls will not clean up at 0.030 inch, the cylinder block must be replaced. Pitted, burned, or nicked valve seats must be reseated. Check the clearances of the valve guides with new valves. If the clearance ex-

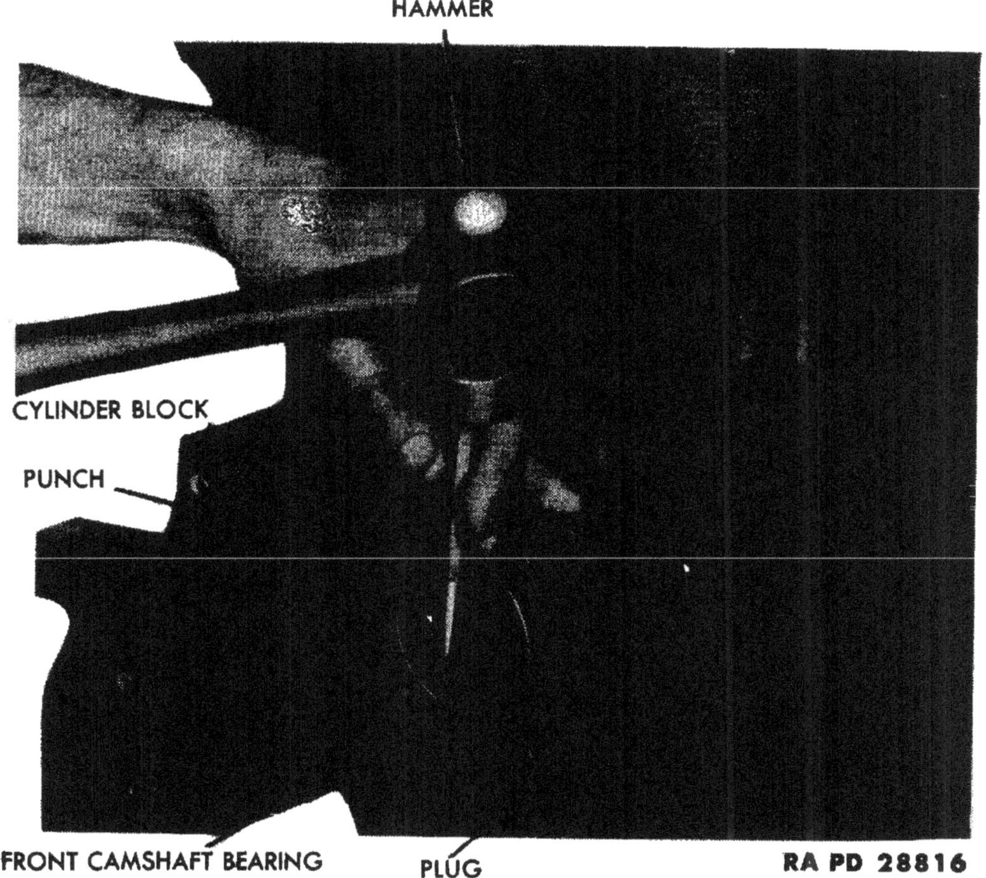

Figure 15 — Staking Camshaft Bearing in Place

ceeds 0.0045 inch in an intake valve guide (using a new intake valve as a gage), or 0.005 inch in an exhaust valve guide (using an exhaust valve as a gage), the valve guides must be replaced (step (6) below). If the clearance exceeds 0.003 inch between valve tappet and valve tappet bore, the valve tappet bores must be reamed to 0.004 inch oversize, and 0.004-inch oversize valve tappets must be installed when assembling engine. If valve tappet bore will not clean up at 0.004 inch oversize, the cylinder block must be replaced.

(4) Replace Studs. Remove all damaged studs with a standard stud puller. To remove a broken stud, indent the end of the broken stud exactly in the center with a center punch. Drill approximately two-thirds through the broken stud with a small drill, then follow up with a larger drill. However, the drill selected must leave a wall thicker than the depth of the threads. Select an extractor (EZ-Out) of the proper size, insert it into the drilled hole, and screw out the

Figure 16 — Removing Water Pump Impeller, Using Puller (41-P-2912)

remaining part of the broken stud. Install the studs with a standard stud driver. Drive all studs until no threads show at the bottom of the studs.

(5) Replace Camshaft Bearing. Drive a punch between the camshaft bearing and cylinder block (fig. 14), and tap the camshaft bearing from the cylinder block. To install the camshaft bearing, drive it in place with a fiber block, making sure the oil hole in the bearing is in line with the oil passage in the cylinder block. Stake the camshaft bearing in place with a punch (fig. 15). Line-ream the camshaft bearing to 2.3145 inches.

(6) Replace Valve Guides. Remove the guides with a suitable valve guide remover. When installing valve guides, drive all intake and exhaust valve guides into the block with a valve guide replacer, leaving a distance of 1 inch from the top of the guide to the top of the cylinder block for exhaust valve guides, and a distance of 1$\frac{5}{16}$ inches for the intake guides.

9. WATER PUMP.

a. **Disassembly.** Pull the water pump bearing retaining wire (fig. 17) from the water pump. Remove the water pump impeller with a puller (41-P-2912) as in figure 16, or press it off in an arbor press. Remove the water pump seal assembly, and water pump seal washer. Press the water pump bearing and shaft assembly, and water pump

ORDNANCE MAINTENANCE — ENGINE AND ENGINE ACCESSORIES FOR ¼-TON 4x4 TRUCK (WILLYS-OVERLAND MODEL MB AND FORD MODEL GPW)

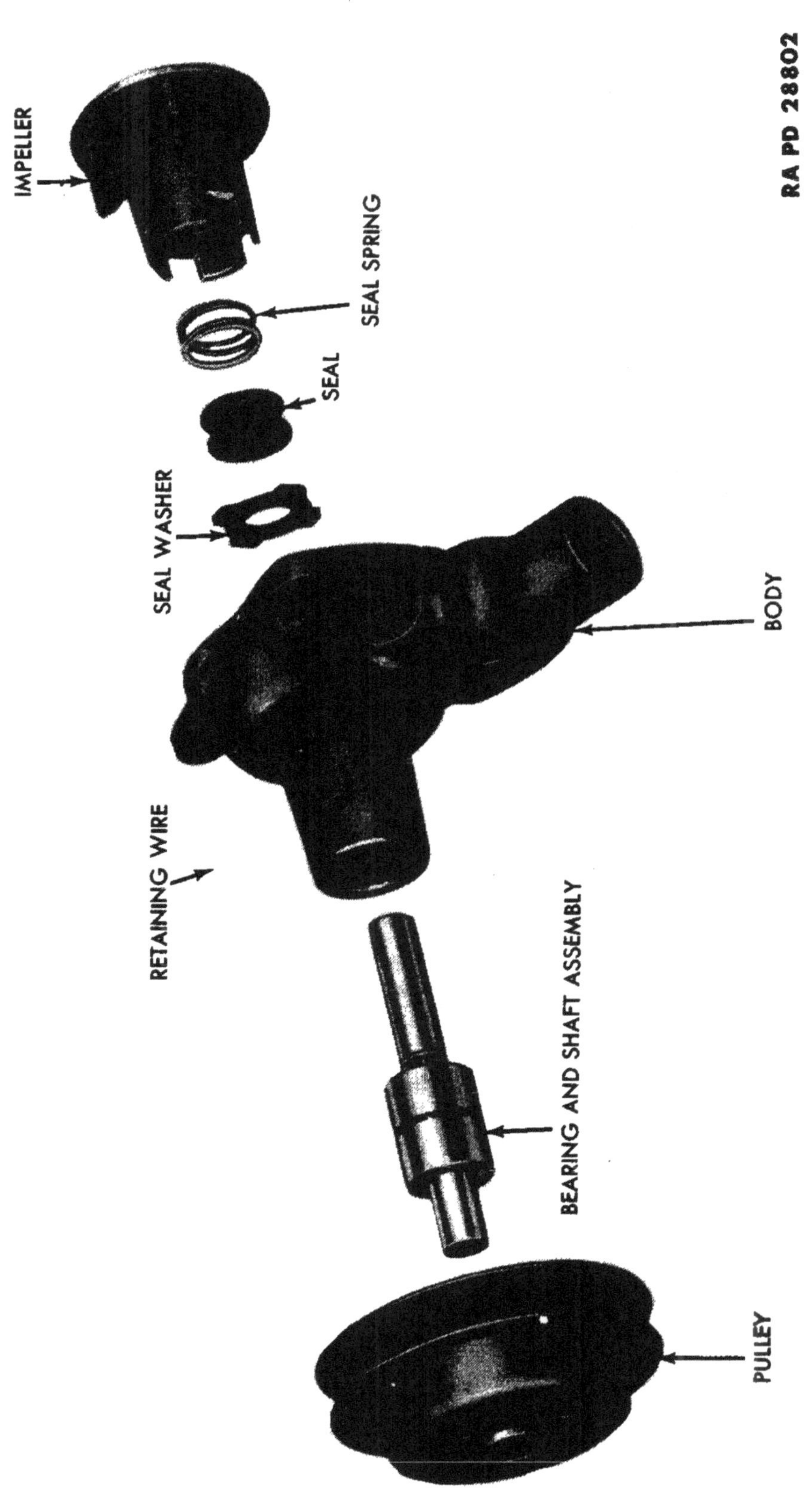

Figure 17 — Water Pump Disassembled

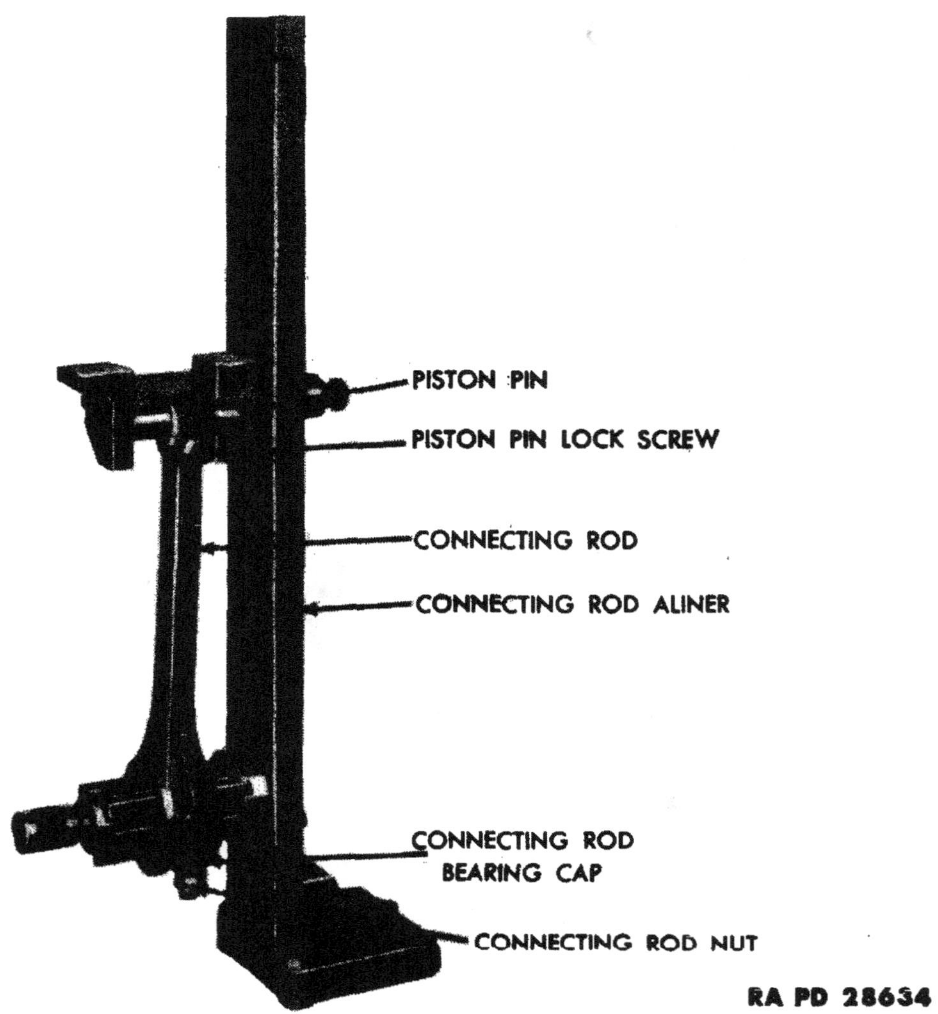

Figure 18 — Checking Connecting Rod Alinement for Twist, using Aliner (41-A-135)

pulley from the water pump body. Press the water pump pulley off the water pump bearing and shaft assembly.

b. **Cleaning.** Clean all parts thoroughly in dry-cleaning solvent.

c. **Inspection and Repair.**

(1) WATER PUMP BODY (fig. 17). A cracked or damaged water pump body must be replaced.

(2) WATER PUMP IMPELLER (fig. 17). A water pump impeller that is cracked or that has a broken fin must be replaced.

(3) WATER PUMP PULLEY (fig. 17). A distorted or damaged water pump pulley must be replaced.

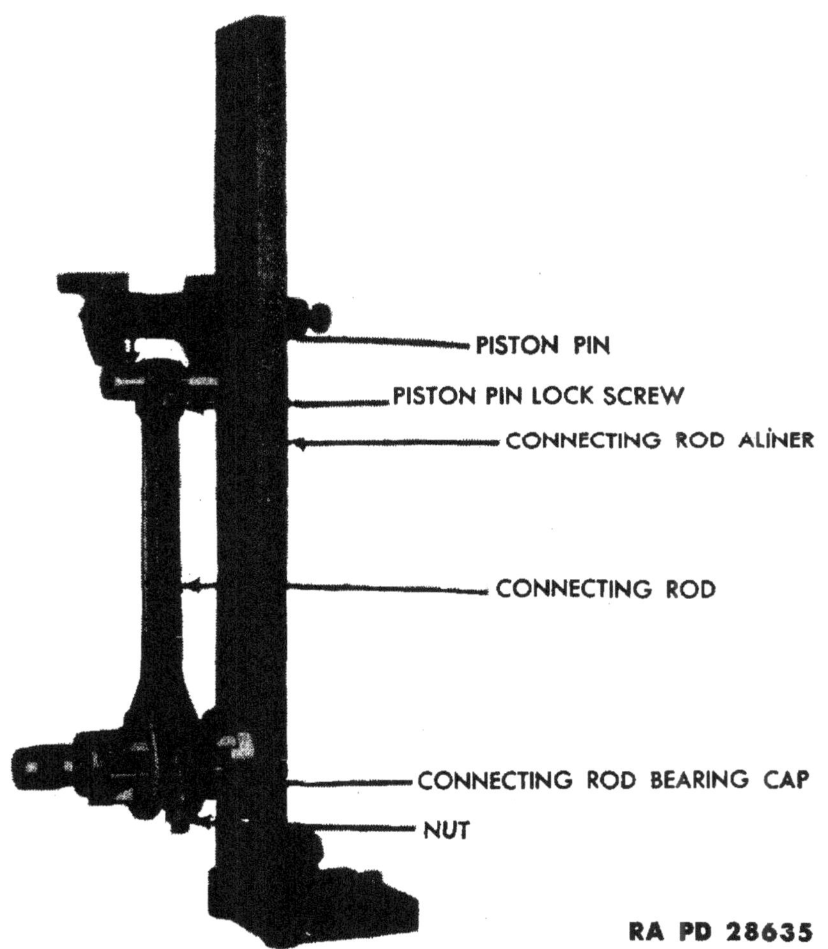

Figure 19 — Checking Connecting Rod Alinement for Bend, Using Aliner (41-A-135)

(4) WATER PUMP BEARING AND SHAFT ASSEMBLY (fig. 17). Rotate the water pump bearing; if the bearing binds or has a tendency to stick, it must be replaced. Bearings that have side or end play must be replaced.

d. **Assembly.** Press the front (short) end of the water pump bearing and shaft assembly into the water pump pulley. Press the water pump pulley and water pump bearing and shaft assembly into the front end of the water pump body until the groove on the bearing is in line with the small slot in the water pump body. Dip a new water pump seal assembly and water pump seal washer in hydraulic brake fluid, and install them in the water pump impeller. Place the impeller in a press, and press the shaft into the impeller until the end of the shaft is flush with the end of the water pump impeller. Install the water pump bearing retaining wire in place.

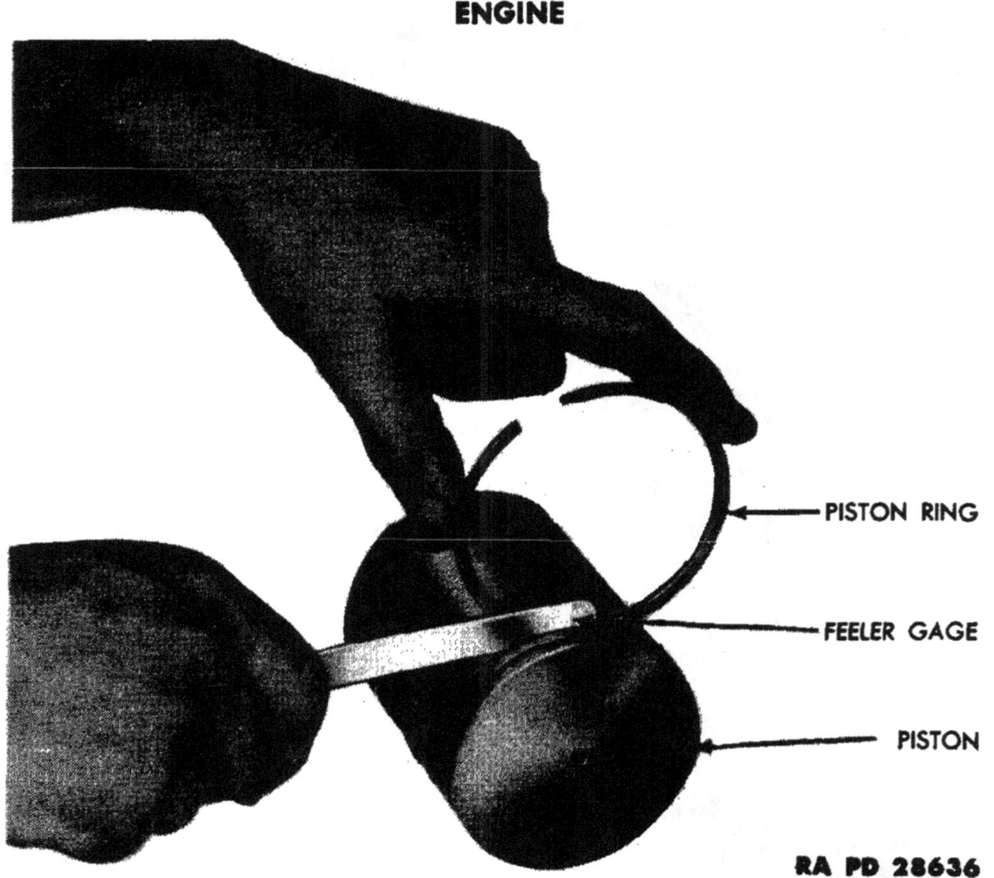

Figure 20 — Checking Clearance of Ring Groove with Feeler Gage

10. CONNECTING ROD AND PISTON ASSEMBLY.

a. **Disassembly.** Remove the piston rings with a standard ring remover. Remove the piston pin lock screw, and push the piston pin out of the piston.

b. **Cleaning.** Scrape the carbon from the ring grooves in the piston, and from the dome. Remove all foreign matter from the oil holes in the oil ring (lower) groove. Clean the complete assembly in dry-cleaning solvent.

c. **Inspection and Repair.** Pistons with cracks, scores, or damage of any kind must be replaced. Determine the wear on the skirt of each piston at the bottom at right angles to the piston pin. If the wear is 0.010 inch less than the original size, or if the piston is out-of-round more than 0.005 inch, the piston must be replaced. Check the width of the ring grooves with new rings and a feeler gage (fig. 20). If the piston ring groove wear exceeds 0.003-inch clearance between the piston ring and ring groove, the piston must be replaced. Measure the piston pin hole. If the inside diameter of the piston pin hole is more than 0.813 inch, the piston must be replaced. Piston pins worn to less than 0.8115-inch

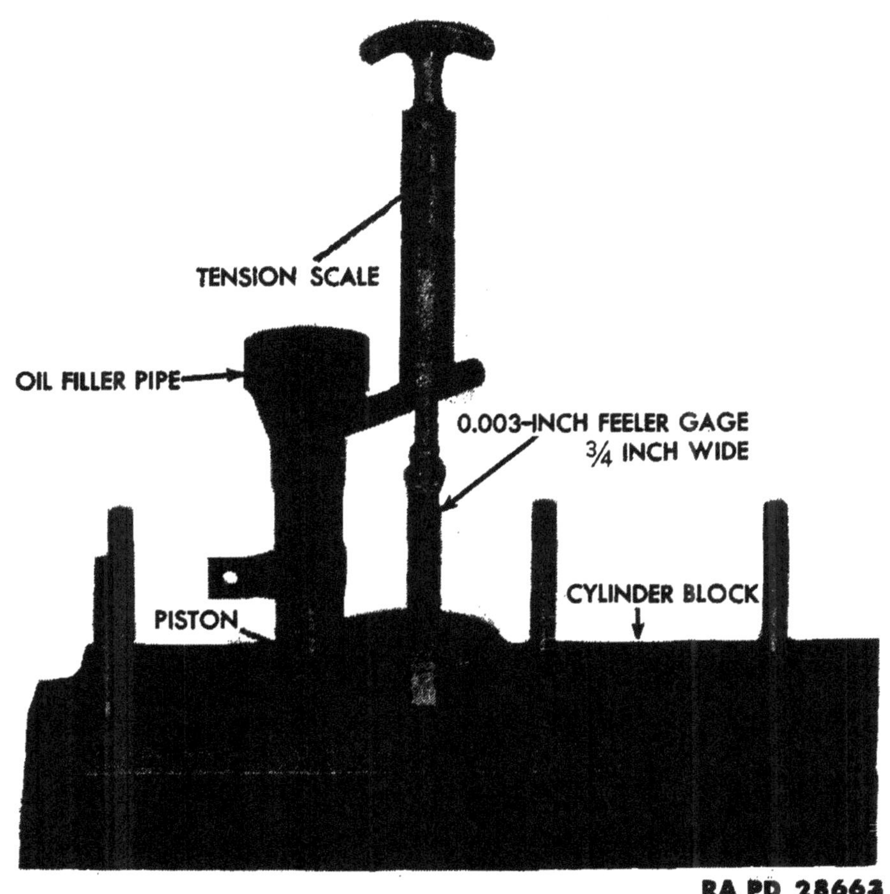

Figure 21 — Fitting Piston in Cylinder Bore, Using Scale w/feelers (41-S-498)

diameter must be replaced. Check the connecting rods for alinement, using aliner (41-A-135) (figs. 18 and 19). Bent or twisted connecting rods must be correctly alined. Damaged connecting rod bolts must be replaced. If connecting rods are fitted with studs, and studs are damaged, the complete connecting rod must be replaced. Excessively worn, scored, discolored, or pitted connecting rod insert bearings must be replaced.

d. **Fit Piston.** The normal clearance of the piston to the cylinder bore is 0.003 inch. Place a piston fitting scale with feelers (41-S-498) into the cylinder bore, making sure the feeler gage is long enough to cover the entire length of a piston. Push a piston into the cylinder bore with the T-slot in the piston opposite the feeler gage (fig. 21). Lift up on the tension scale; if more than 10 pounds is required to pull the feeler gage from the cylinder bore, the piston is too tight. Select a

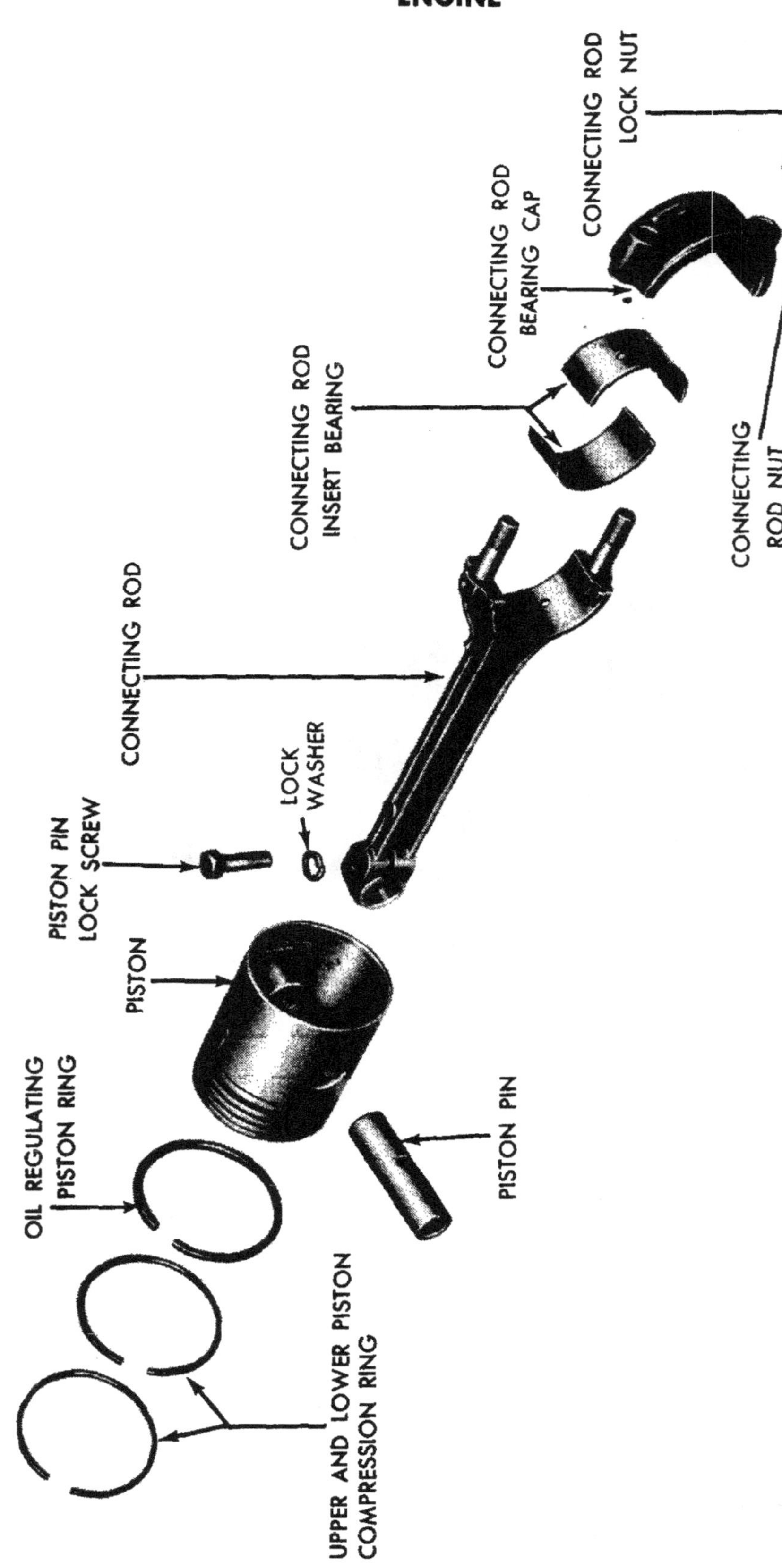

Figure 22 — Connecting Rod and Piston Assembly, Disassembled

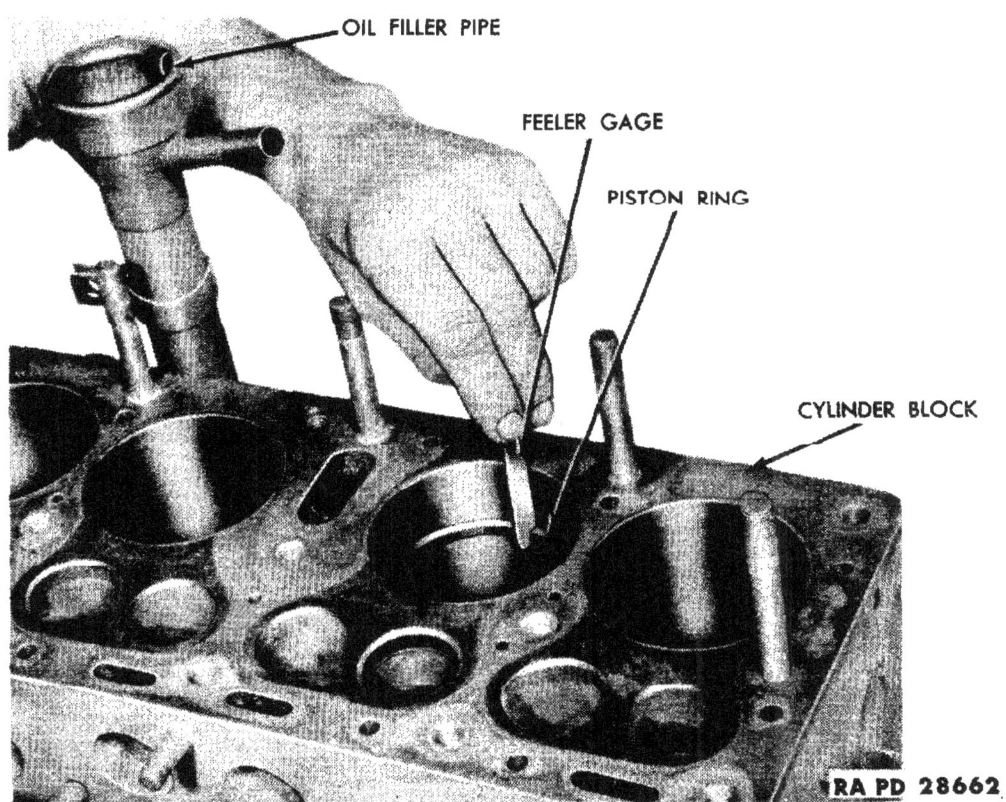

Figure 23 — Measuring Piston Ring End Gap with Feeler Gage

smaller piston. If less than 5 pounds pull is required to remove the gage, the piston is too loose. Select a larger piston. Mark the cylinder number on each piston after fitting.

e. **Assemble Piston, Piston Pin, and Connecting Rod.** When installing connecting rods on pistons, make sure the oil squirt hole in the connecting rod is opposite the T-slot in the piston (fig. 36). If assembled in this manner, the off-set on the connecting rods will be in the correct position when installed in the cylinder block (par. 18 **f**). Select a piston pin which can be inserted in the piston with a light "push" fit (piston temperature at 70° **F**), and push it part way into the piston pin hole, with the groove in the piston pin facing downward. Hold the connecting rod in line with the piston pin hole, and push the piston pin the rest of the way into the piston. Install and tighten the piston pin lock screw in the connecting rod.

f. **Fit and Install Piston Rings.** Place a new piston ring in the cylinder bore, and press it about halfway down into the cylinder bore with the bottom of a piston, so that the ring will be square with the cylinder wall. Measure the piston ring end gap with a feeler gage (fig.

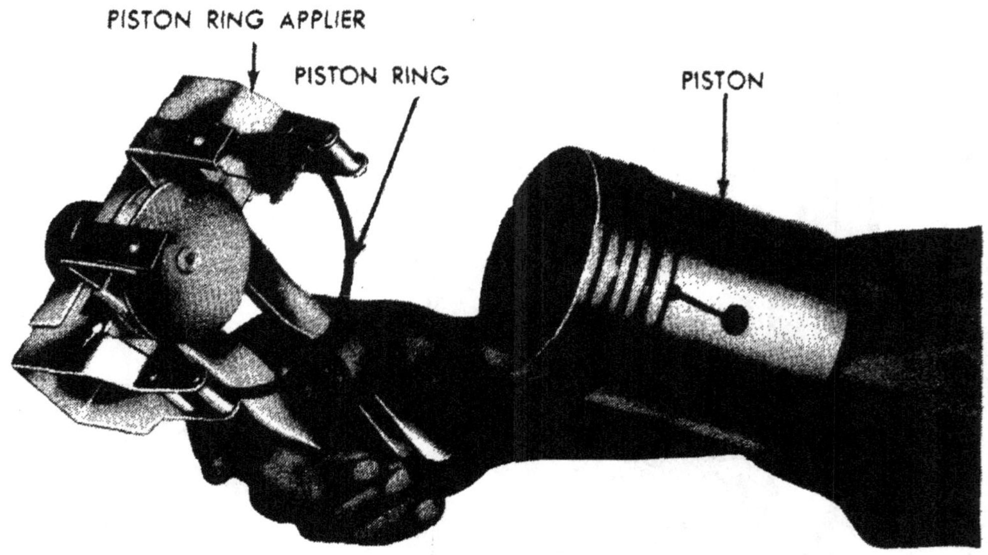

Figure 24 — Installing Piston Ring on Piston, Using Applier (41-A-329-500)

23). If the gap is less than 0.008 inch, remove the ring, and file with a fine-cut file until the correct gap (0.008 to 0.013 inch) is obtained. If end gap exceeds 0.013 inch, an oversize ring must be used. Repeat the same procedure for all piston rings. Roll the new piston ring around its particular groove in the piston. The ring should roll freely, and not have a clearance of more than 0.003 inch (fig. 20). Repeat the same procedure on each piston ring. Install the piston rings on the piston with a piston ring applier (41-A-329-500) (fig. 24), making sure that the beveled edge of both compression rings are towards the top.

11. CAMSHAFT ASSEMBLY.

a. **Cleaning.** Clean the camshaft, camshaft sprocket, camshaft thrust washer, and camshaft thrust spring and plunger, in dry-cleaning solvent.

b. **Inspection and Repair.** A camshaft with excessively scored or damaged cams, or with worn, corroded, scored, or discolored journals, must be replaced. Inspect the camshaft oil pump drive gear. If the teeth are worn, broken, or chipped, the camshaft must be replaced. Measure the four camshaft journals (fig. 25), and record the readings. If reading is less than 2.185 inches for the front journal, 2.122 inches for the front intermediate journal, 2.0595 inches for the rear intermediate journal, and 1.622 inches for the rear journal, the camshaft must be replaced. A camshaft gear with worn, broken, or chipped teeth must be replaced. Small nicks can be honed, and then polished

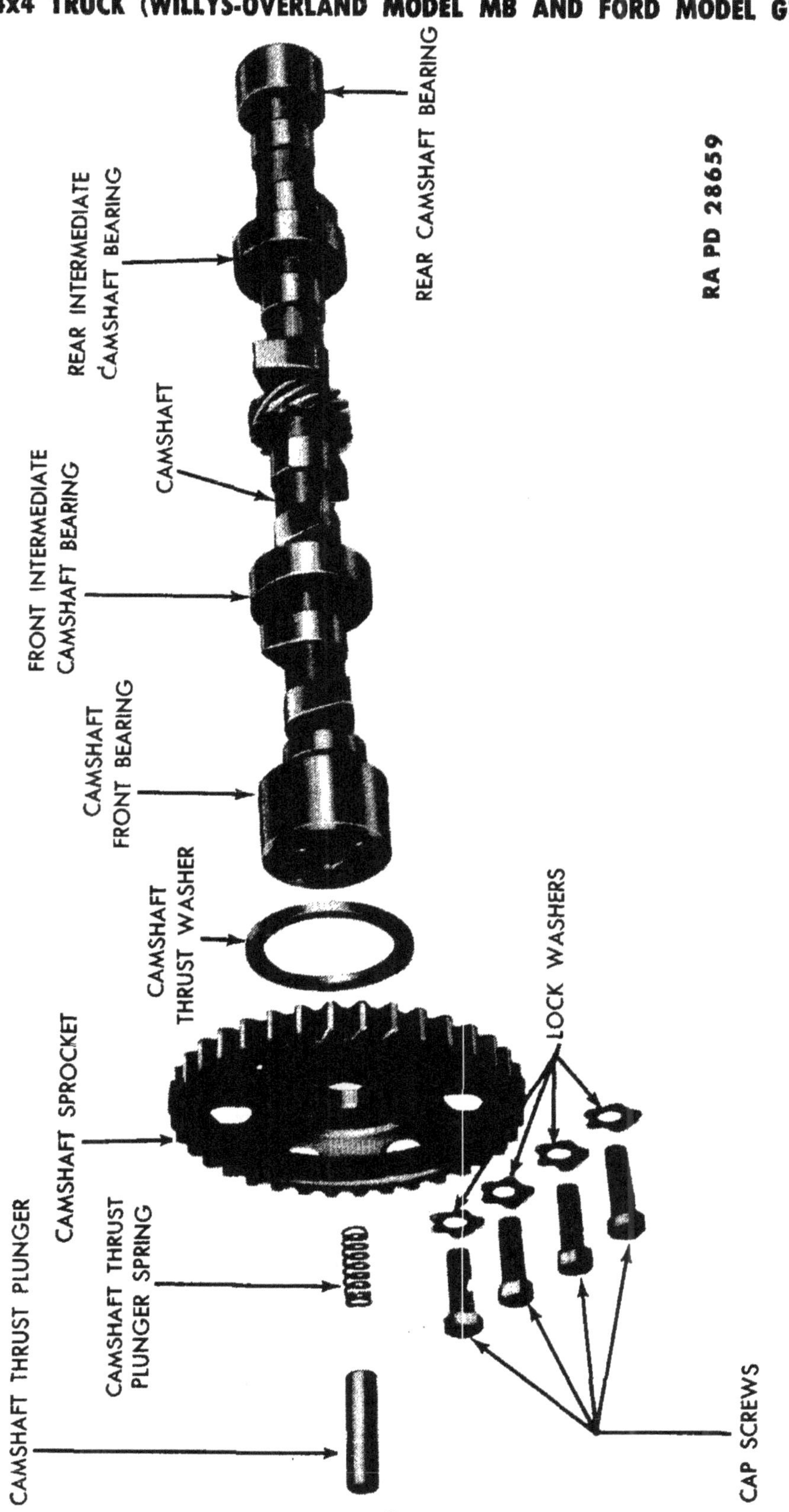

Figure 25 — Camshaft Assembly Disassembled

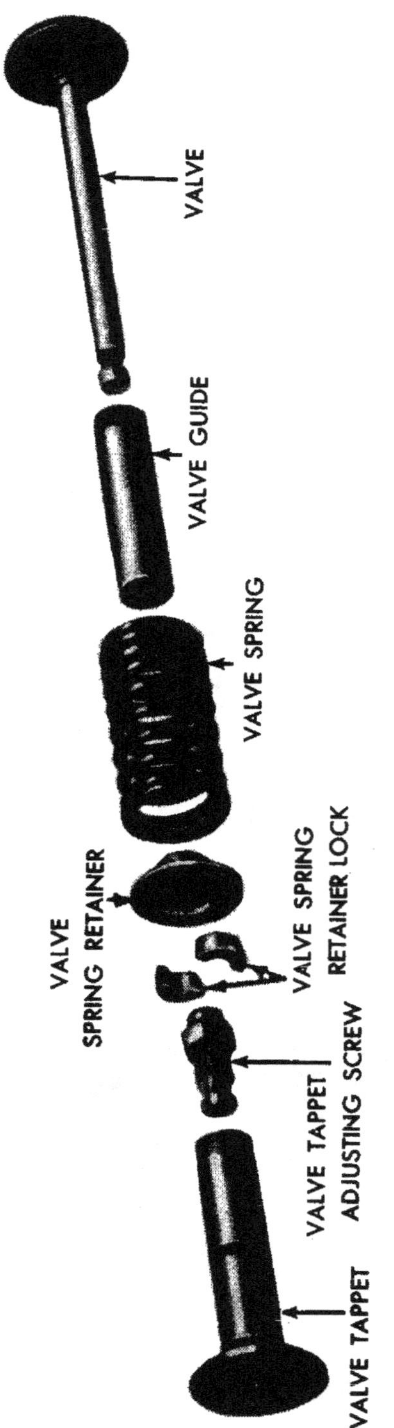

Figure 26 — Valve Assembly Disassembled

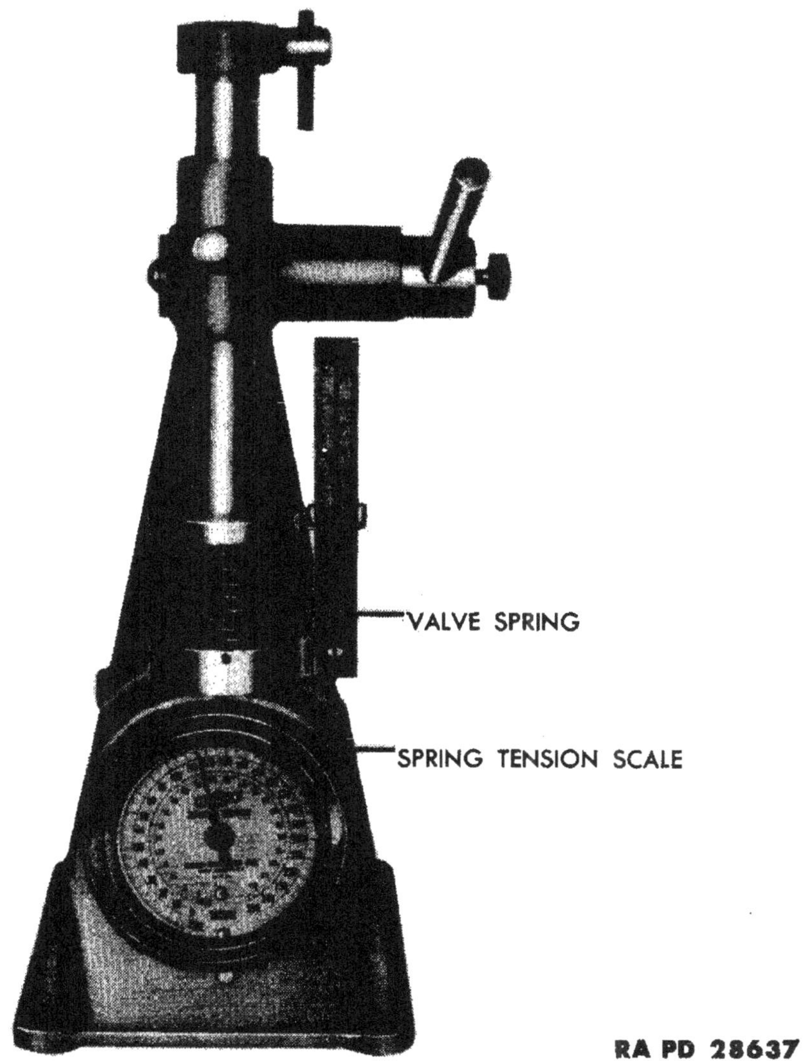

Figure 27 — Checking Tension of Valve Spring, Using Tester (41-T-1600)

with a fine stone. A weak (less than 15 pounds compressed to $^{29}/_{32}$ inch) or broken camshaft thrust plunger spring must be replaced.

12. VALVE AND VALVE SPRINGS.

a. Cleaning. Scrape the carbon off the valve heads and stems. Clean the valves and valve springs thoroughly in dry-cleaning solvent.

b. Inspection and Repair. Valves with bent or scored stems must be replaced. Measure the outside diameter of each valve stem (fig. 26). If measurement is less than 0.3685 inch for the exhaust valve, or 0.368 inch for the intake valve, the valves must be replaced. Pitted, corroded, or burned valves must be refaced. Valves that are burned, warped, or pitted, and will not clean up with a light cut of the grinding wheel, must be replaced. Measure the free length of each valve spring; if less than 2½ inches in length, the spring must be replaced. Check the tension of each valve spring (fig. 27), using tester (41-T-1600). If the valve spring registers less than 50 pounds when compressed to $2\frac{1}{16}$ inches, or 116 pounds when compressed to 1¾ inches in length, it must be replaced.

13. VALVE TAPPETS.

a. Cleaning. Clean the valve tappets thoroughly in dry-cleaning solvent.

b. Inspection and Repair. Cracked, scored, or excessively worn valve tappets (fig. 26) must be replaced. Valve tappets, or valve tappet adjusting screws (fig. 26) with worn or damaged threads, must be replaced.

c. Disassembly. Unscrew the valve tappet adjusting screw from the tappet.

d. Assembly. Screw the valve tappet adjusting screw approximately three-quarters of the way into the valve tappet.

14. OIL PUMP AND OIL INTAKE FLOAT.

a. Disassembly. Remove the screw that holds the oil pump cover assembly to the oil pump, and remove the cover. Remove the oil pump relief spring retainer, gasket, shims, spring, and plunger from the oil pump cover (fig. 28). File either side of the oil pump driven gear pin (fig. 28), until the pin is flush with the driven gear sleeve. Drive the pin out of the sleeve and shaft with a small punch. Pull the oil pump shaft assembly out of the housing. Remove the cotter pin that holds the intake oil float to the oil float support, and remove the float (fig. 11). Straighten the four tabs on the oil intake float sump, and remove the sump. Lift the screen from the oil intake float.

b. Cleaning. Clean all parts and drilled passages thoroughly with dry-cleaning solvent, and blow out the oil intake float screen and all oil passages in the oil pump and oil intake float.

c. Inspection and Repair. A cracked or damaged oil pump housing or cover must be replaced. Measure the small pinion shaft on the oil pump cover. If less than 0.372 inch, the cover must be replaced. Measure the inside diameter of the oil pump housing (shaft end) (fig. 28). If larger than 0.505 inch, the oil pump housing must be replaced. An oil pump shaft assembly with broken teeth, or with a shaft measuring under 0.495 inch, must be replaced. An oil pump shaft assembly with a distributor slot worn more than three-sixteenths inch, must be

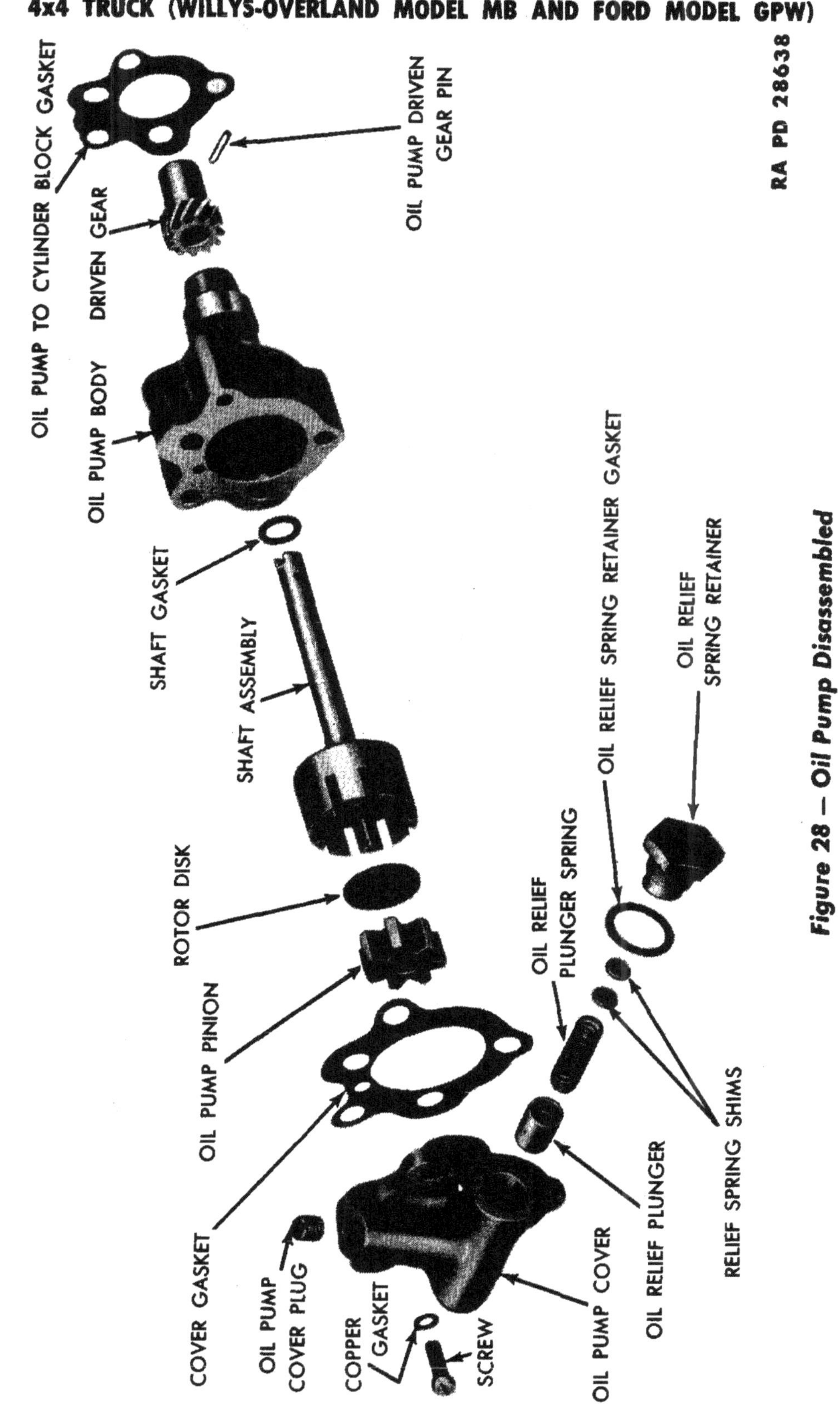

Figure 28 — Oil Pump Disassembled

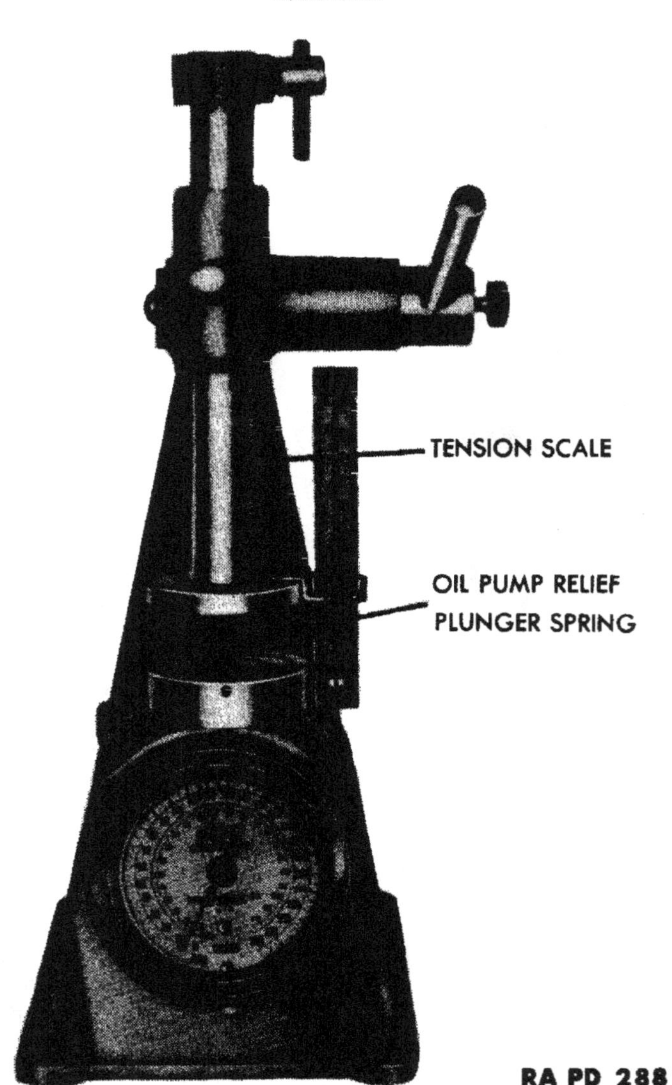

Figure 29 — Checking Oil Pump Relief Valve Spring Tension, Using Tester (41-T-1600)

replaced. An oil pump pinion gear with broken or worn teeth, or with an inside diameter of more than 0.378 inch, must be replaced. Measure the rotor disk (fig. 28); if less than 0.069 inch thick, it must be replaced. An oil pump driven gear with broken or chipped teeth must be replaced. Compress the oil pump relief valve spring to $1\frac{1}{16}$ inches (fig. 29), using tester (41-T-1600). If the tension is less than $5\frac{1}{2}$ pounds, the spring must be replaced. Replace a broken or cracked oil intake float support; also a distorted or leaking intake float support.

d. Assembly. Place the screen in the oil intake float. Place the sump on the oil intake float, and bend the four tabs to lock the sump to the float. Slide the oil intake float support onto the float, making sure the tongue on the support is in the recess. Install a cotter pin in the support. Slide a new oil pump shaft gasket on the shaft assembly. Slide the shaft assembly into the oil pump housing. Tap the driven gear onto the shaft with the gear toward the oil pump, until there is 0.0312-inch clearance between the gear and oil pump body. If installing a new shaft, drill a hole for the pin, and install a new driven gear pin through the gear and shaft. Peen both ends of the driven gear pin. Install the rotor disk in the shaft assembly. Install the pinion gear on the oil pump cover. The pinion gear must have from 0.001- to 0.003-inch end play, measured from the end of the pinion shaft. Place a new gasket on the oil pump cover, and install the cover onto the housing. Install the copper gasket and hold-down screw in the cover. Tap the oil pump shaft into the housing, and check the clearance between the driven gear and housing. Insert a screwdriver between the gear and housing, and pry on the shaft. Remove screwdriver and again measure clearance. The difference represents the end play, and must be 0.002- to 0.004 inch. If sufficient, remove cover and add sufficient gaskets. Drop the oil relief plunger and spring (fig. 28) into the opening in the oil pump cover. Place two oil relief spring shims into the oil relief spring retainer (fig. 28). Place a new gasket on the oil relief spring retainer, and install and tighten the retainer to the cover. Install and tighten the oil pump cover plug.

15. CRANKSHAFT ASSEMBLY.

a. **Cleaning.** Clean out the drilled holes on the crankshaft journals with a piece of wire. Clean the crankshaft thoroughly with dry-cleaning solvent.

b. **Inspection and Repair.** Inspect all crankshaft journals. If worn or scored, the crankshaft must be replaced or reworked. Measure the outside diameter of each crankshaft journal. If the diameter is less than 1.9365 inches on the crankpin journals (fig. 30), or 2.3325 inches on the main bearing journals (fig. 30), or if any of the journals are out-of-round more than 0.0005 inch, the crankshaft must be reworked to 0.010-, 0.020-, or 0.030-inch undersize, whichever the case may be. Light scores and scratches can be honed, and then polished with crocus cloth. Crankshafts that will not clean up at 0.030-inch undersize must be replaced If a new crankshaft or flywheel is being used, it must be fitted as outlined in paragraph 16 e.

c. **Remove Crankshaft Sprocket** (fig. 30). Install a standard puller on the crankshaft sprocket and remove the sprocket. Remove the Woodruff key, spacer, thrust washer, and shims.

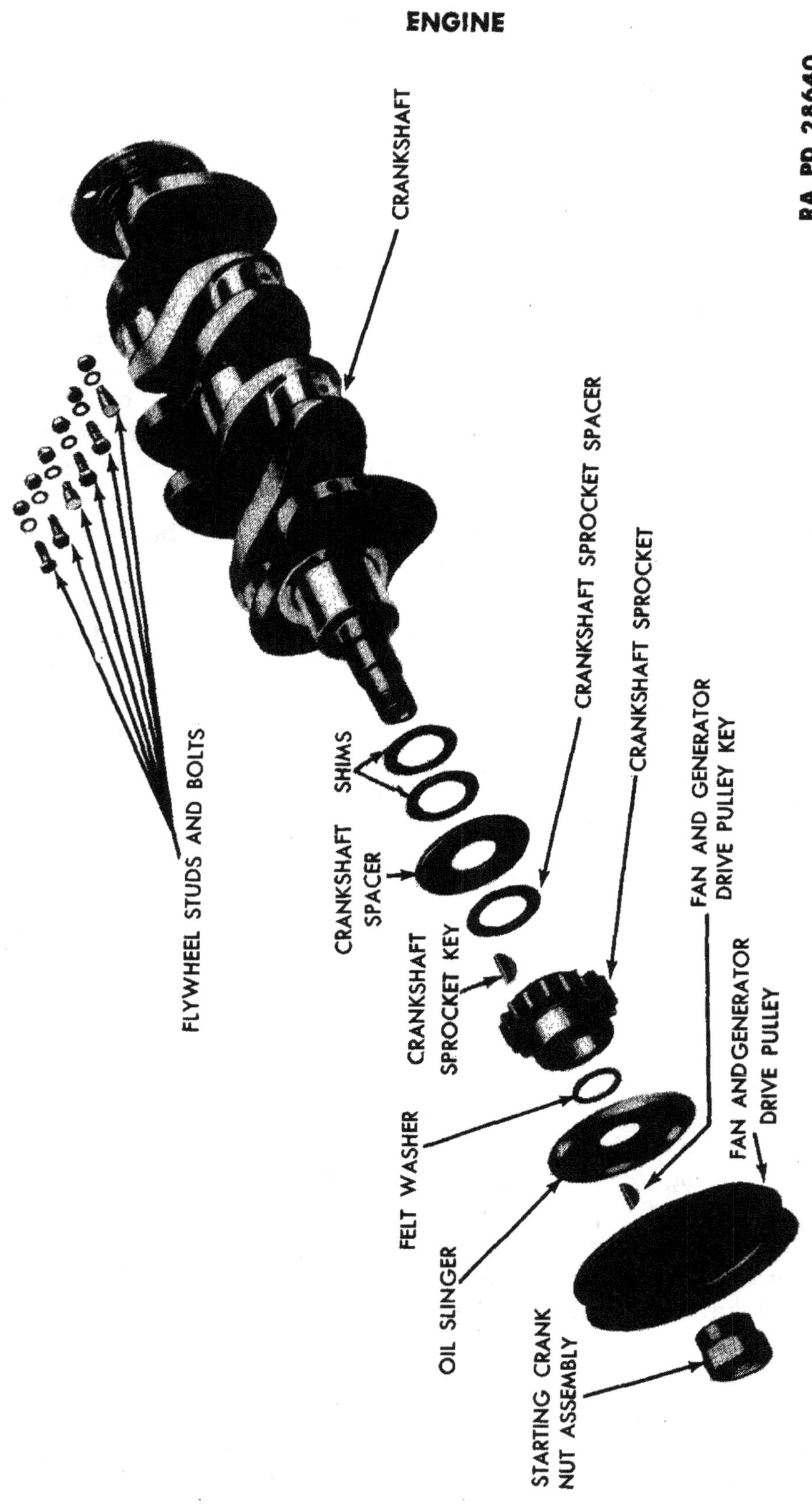

Figure 30 — Crankshaft Assembly Disassembled

Figure 31 — Flywheel Ring Gear and Pilot Bushing

16. FLYWHEEL ASSEMBLY.

a. **Cleaning.** Wash the flywheel thoroughly in dry-cleaning solvent.

b. **Inspection and Repair.** A flywheel (fig. 31) with an excessively scored or worn friction face must be replaced. A flywheel ring gear with broken, chipped, or excessively worn teeth must be replaced (subpars. c and d below). Measure the inside diameter of the main drive gear pilot bushing. If more than 0.632 inch, it must be replaced (subpars. c and d below). If a new crankshaft or flywheel is being used, it must be fitted as outlined in subparagraph e below.

c. **Disassembly.** Drive the main drive gear pilot bushing out of the flywheel. Heat the flywheel ring gear until it can be driven off the flywheel.

d. **Assembly.** Clean the flywheel ring gear recess on the flywheel. Apply heat evenly to the ring gear. When the ring gear is thoroughly heated, place it on the cold flywheel, making sure it is firmly seated in its recess. Drive a main drive gear pilot bushing in place with a fiber block.

e. **Fit Crankshaft to Flywheel When Either Part Is New.** Install the flywheel onto the crankshaft with the four crankshaft bolts, lock washers and nuts, making sure the index mark on the crankshaft is in line with the index mark on the flywheel. Drill the two tapered (stud) holes with a $^{35}/_{64}$-inch drill, and ream the two holes with a $^{9}/_{16}$-inch (0.5625-inch) reamer. Install the two bolts that are supplied with each crankshaft and/or flywheel.

17. INTAKE AND EXHAUST MANIFOLDS.

a. **Disassembly.** Remove the four cap screws that secure the intake manifold to the exhaust manifold, and separate the two manifolds. Remove the nut and bolt that hold the heat control valve shaft. Pull the counterweight lever, heat control lever, washer, and spring off the shaft.

b. **Cleaning.** Scrape all the old gaskets and carbon from the manifolds. Wash the manifold and parts in dry-cleaning solvent.

c. **Inspection and Repair.** Cracked or broken manifolds must be replaced. Damaged or broken studs must be replaced. An exhaust manifold with a damaged exhaust valve control or shaft must be replaced (par. 7 b).

d. **Assembly.** Slide the heat control valve spring onto the shaft, making sure the end of the heat control valve spring is resting on top of the stop. Slide the washer, counterweight, and control lever onto the shaft; install the nut and bolt through the counterweight. Place a new gasket between the two manifolds, and install the four cap screws and heat control spring stop.

Section V

ASSEMBLY OF ENGINE

18. ASSEMBLY.

a. **Install Valves.** Place a valve tappet in each valve tappet bore. Slide the camshaft into the cylinder block. Install a valve spring and valve spring retainer on each tappet, making sure the closed coils of the valve springs are against the cylinder block. Install the valves

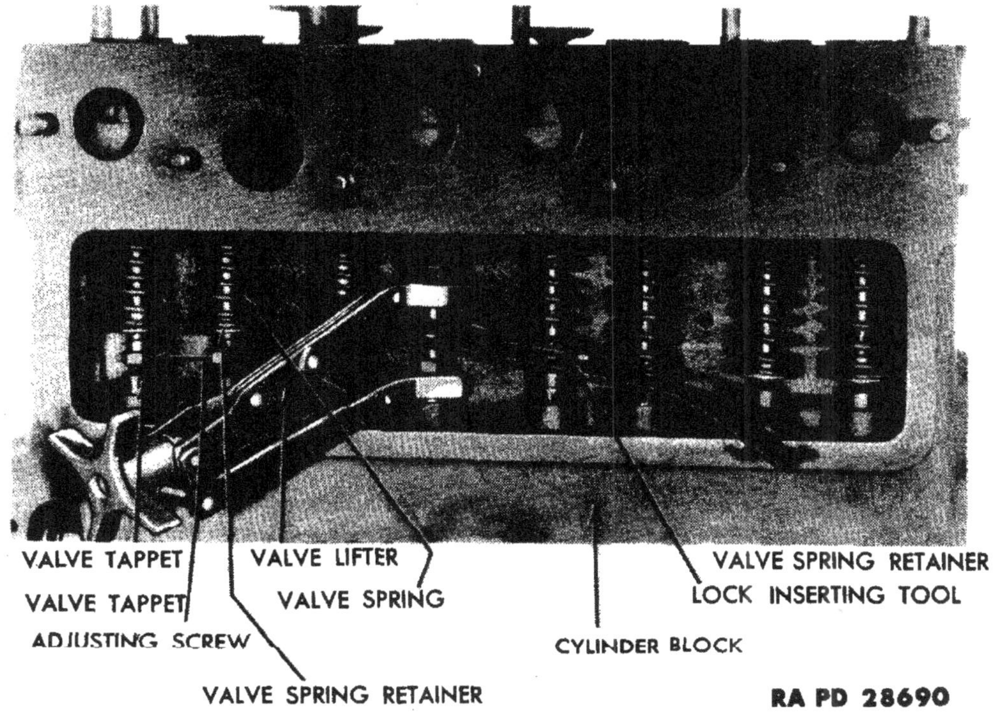

Figure 32 — Installing Valve Spring Retainer Locks, Using Lifter (41-L-1410) and Replacer (41-R-2398)

in their respective valve guides. Compress the valve springs on all valves that are in closed position using valve lifter (41-L-1410), and install the lower valve spring retainer locks (fig. 32), using replacer (41-R-2398). Turn the camshaft to close the other valves, and install the lower valve spring retainer locks on the rest of the valves.

b. Adjust Valve Tappets. Turn the camshaft until No. 1 valve is in a closed position, and the tappet is on the heel of the cam. Hold the valve tappet with one wrench, and turn the valve tappet adjusting screw with another wrench (fig. 33) clockwise or counterclockwise until 0.014-inch clearance is established between the valve and the valve tappet adjusting screw. Repeat the same procedure on each valve.

c. Install Crankshaft. If a new crankshaft or flywheel is being used, refer to paragraph 16 e. Install the three upper halves of the main bearing inserts in the cylinder block (fig. 13). Press the rear main bearing crankshaft packing into the recess provided at the rear main bearing (fig. 13), and in the rear main bearing cap. Cut the ends of the crankshaft packing flush with the crankcase and with the bearing cap. Install the four bolts and the two tapered studs in the

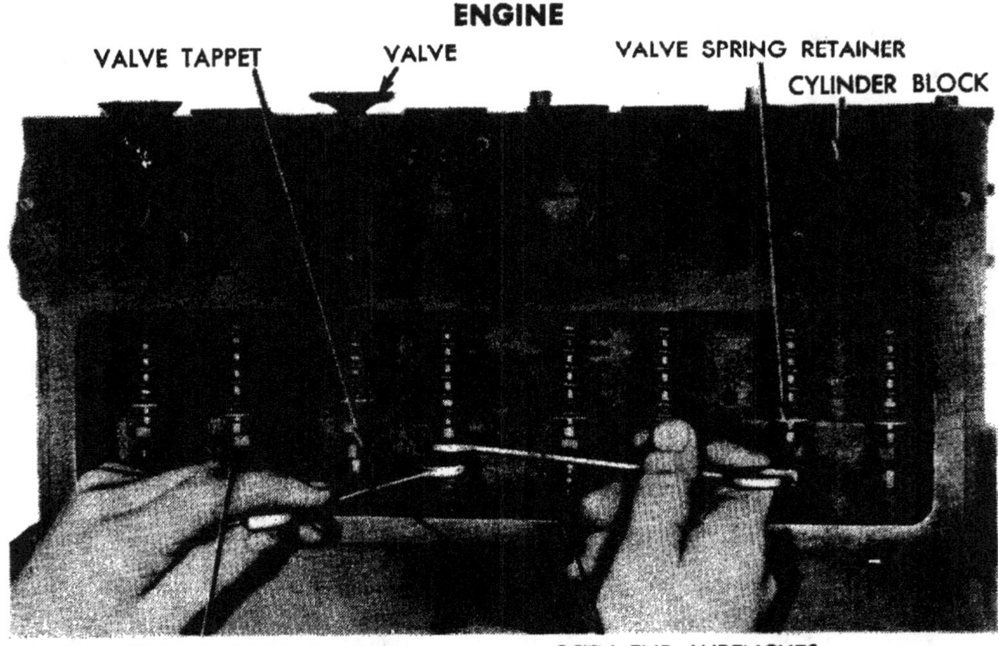

Figure 33 — Adjusting Valve Tappers, Using Wrenches (41-W-3575)

flywheel flange on the crankshaft. Install the three lower halves of the main bearing inserts in the three main bearing caps. Oil the main bearing inserts with a light oil. Place the crankshaft in place in the cylinder block. Install the front and center bearing caps, and tighten the bolts until they are just snug. Coat the rear bearing cap with joint and thread compound on both sides and top. Install the rear bearing cap in the cylinder block. Tighten the six main bearing bolts with a torque wrench to from 65 to 70 foot-pounds. Slip the rear bearing cap packing into the hole on each side of the rear main bearing cap, leaving ¼ inch of the packing to protrude from the crankcase.

d. **Fit Crankshaft.** Place a 0.006-inch feeler gage between the front main bearing cap and the crankshaft, and pull the crankshaft toward the front of the engine as far as possible. Place a straightedge across the front main bearing, and measure the distance between the straightedge and crankshaft to determine the amount of shims to be used (fig. 34).

e. **Check Crankshaft End Play.** Install the necessary amount of shims on the crankshaft to take up the space between the straightedge and crankshaft (fig. 35). Install the crankshaft thrust washer and spacer washer (fig. 30). Tap the large Woodruff key in the crankshaft, and slide the crankshaft sprocket, felt, and crankshaft oil slinger (fig. 30) on the shaft. Tap the small Woodruff key in the crankshaft, install the generator and fan belt drive pulley and cranking nut,

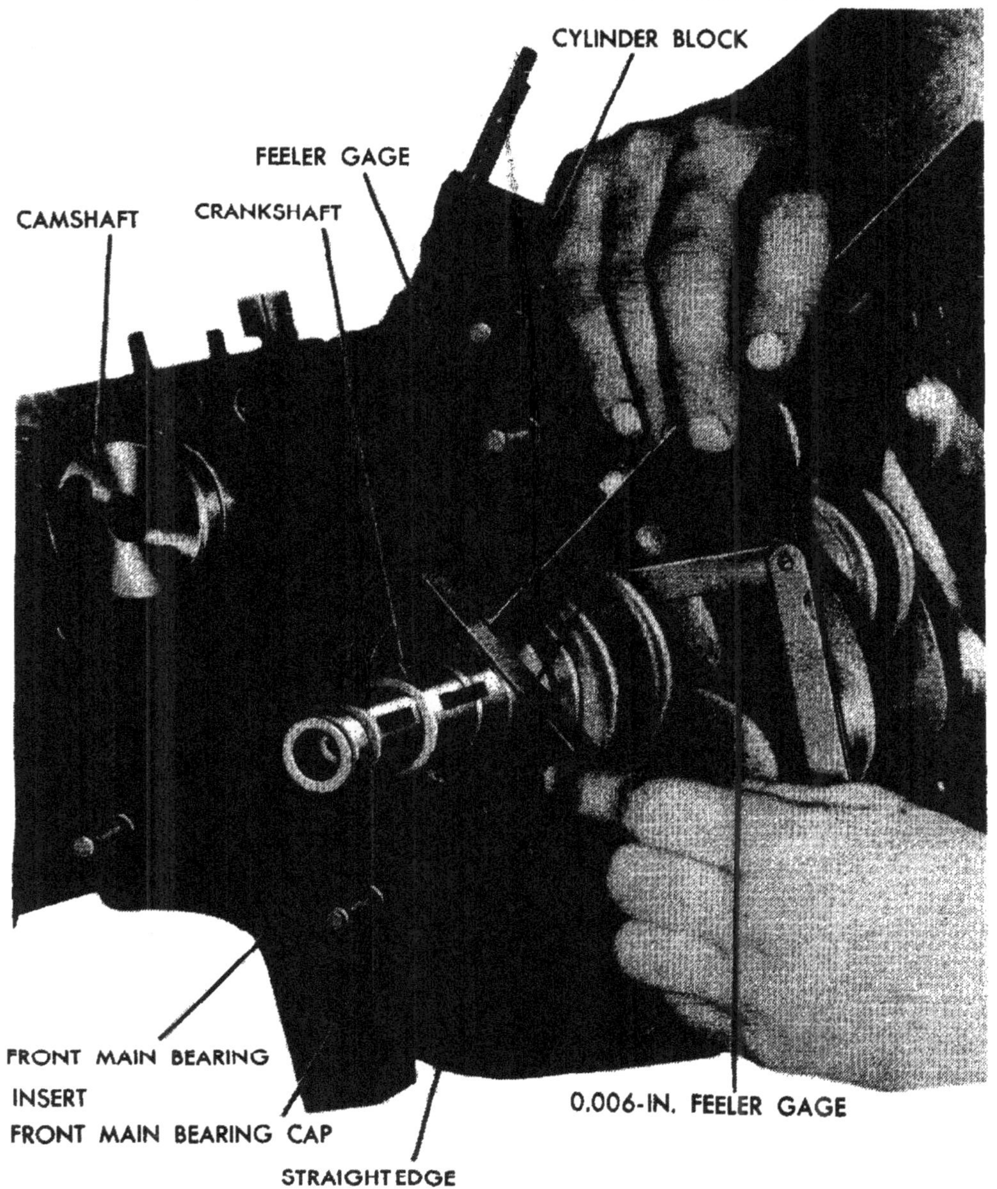

RA PD 28695

Figure 34 — Measuring Crankshaft End Play

and tighten the cranking nut. Place a feeler gage between the front main bearing cap and crankshaft. If more than 0.006-inch end play exists, shims must be removed. If less than 0.004 inch, shims must be added (fig. 35).

f. Install Connecting Rod and Piston Assemblies. (Piston assemblies will have previously been selected for each cylinder as outlined in paragraph 10 d). Oil the piston rings and install a ring compres-

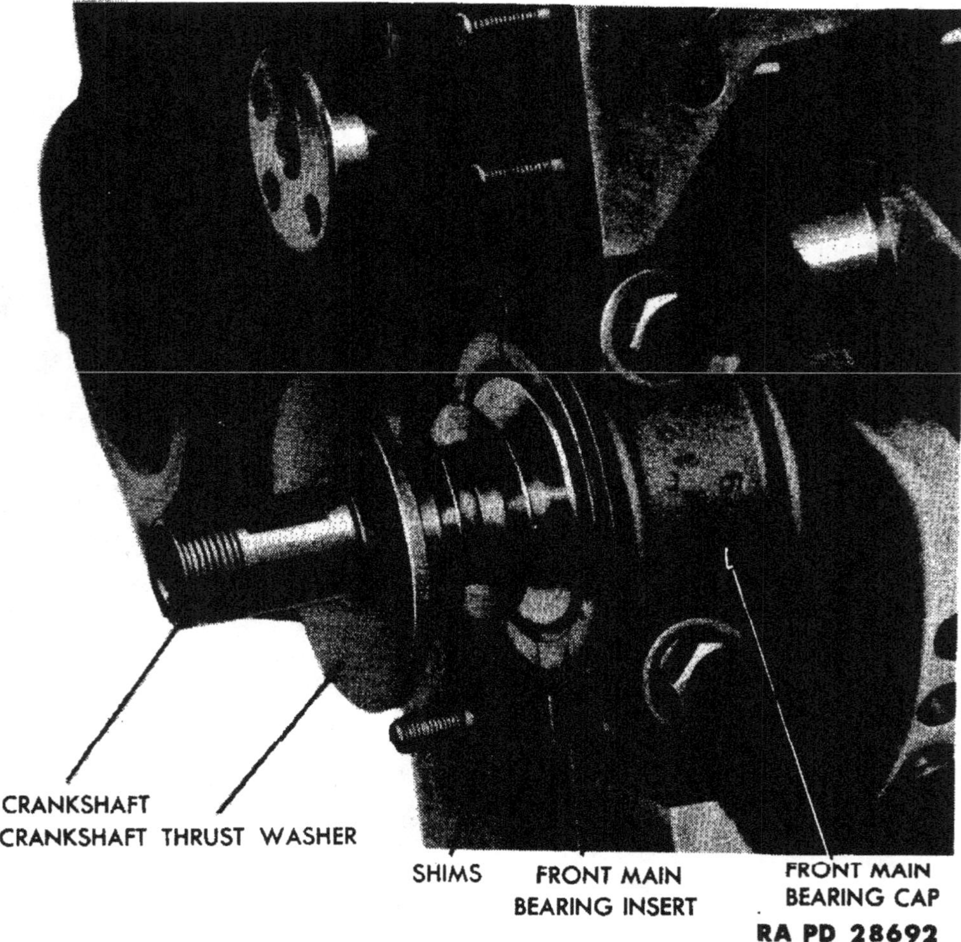

Figure 35 — Shims in Place on Crankshaft

sor (41-C-2550) on the piston rings. Place the No. 1 connecting rod and piston assembly in the No. 1 cylinder with the offset on the connecting rod away from the nearest main bearing (fig. 36). With the T-slot of the piston to the left, and the oil squirt hole in the connecting rod facing toward the right-hand side of the engine, tap the piston down into the cylinder with the handle end of a hammer (fig. 37). Place one-half of a connecting rod insert bearing in the connecting rod, and the other half in the connecting rod bearing cap. Coat the connecting rod insert bearings with a light film of oil. Connect the rod to the crankshaft and install, but do not tighten, the two connecting rod nuts. Repeat the same procedure when installing the other rods, making sure the offset on each connecting rod is away from the nearest main bearing, and the oil squirt hole facing toward the left-hand side of the engine. Tighten all the connecting rod nuts to from 50 to 55 foot-pounds pull with a torque wrench. Install a

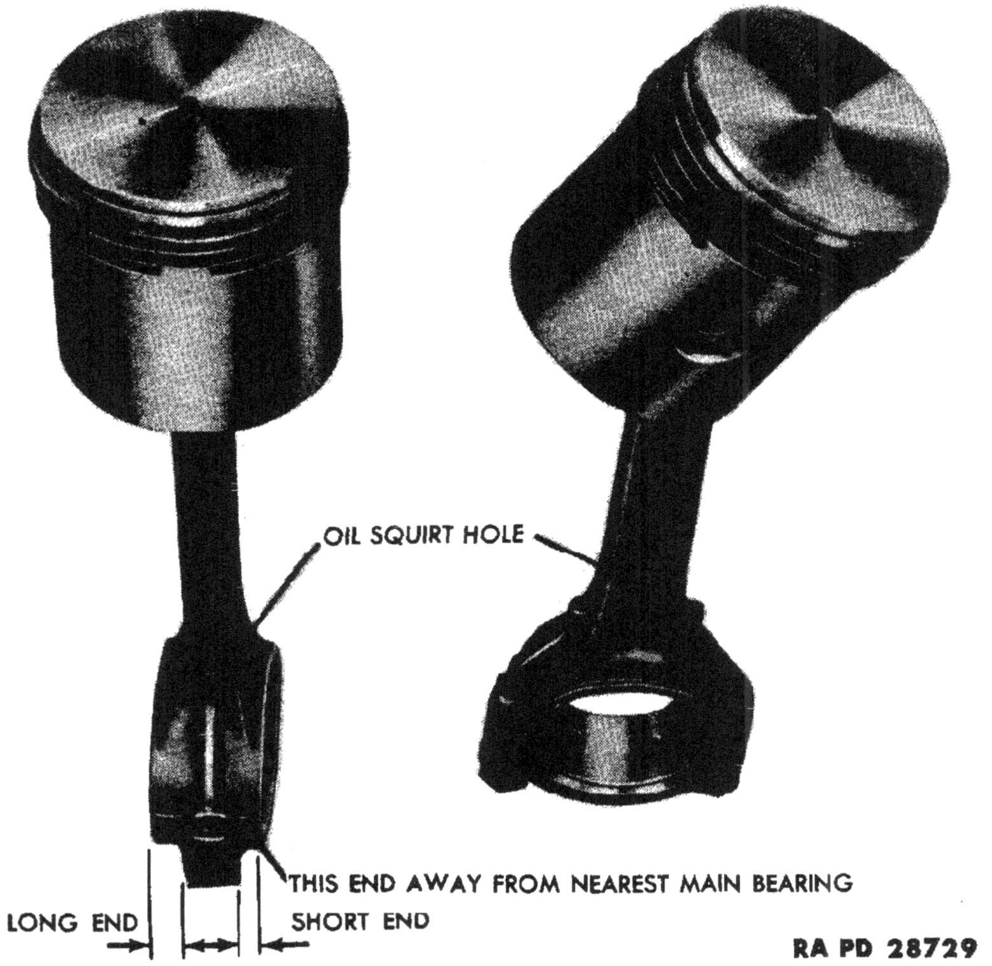

Figure 36 — Position of Connecting Rod Off-set and Oil Squirt Hole When Installed In Engine

pal nut on each connecting rod stud or bolt. Turn the pal nuts down on the stud or bolt until sealed, then turn one complete turn.

g. **Install Flywheel.** If installing a new flywheel or crankshaft, fit the crankshaft to the flywheel as outlined in paragraph 16 e. Fasten the engine rear plate temporarily to the engine with two bolts. Turn the crankshaft until the No. 1 and No. 4 pistons are at top center. Place the flywheel on the crankshaft flange so that the letters "TC" on the flywheel are lined up with the index mark at the center of the timing hole (fig. 38) in the engine rear plate, and the index mark on the crankshaft flange and on the flywheel are in line with each other. Install and tighten the six lock washers and nuts on the flywheel from 36 to 40 foot-pounds with a torque wrench. Check run-

Figure 37 — Installing Piston and Connecting Rod Assembly in Cylinder Block, Using Ring Compressor (41-C-2550)

out on the flywheel with a dial gage. If the run-out exceeds 0.008 inch at the outer edge, the flywheel or crankshaft flange must be refaced.

h. Install Clutch Disk and Pressure Plate. Hold the clutch disk on the flywheel, and install a clutch pilot tool in the flywheel and the disk. Hold the pressure plate on the flywheel and install, but do not tighten, six lock washers and cap screws (fig. 39). Tighten the six cap screws evenly to prevent bending the pressure plate frame. Remove the clutch pilot.

i. Install Camshaft Sprocket. Place a gasket and the engine front plate on the engine, and install the three cap screws. Turn the crankshaft until No. 1 piston is at top center (fig. 38). Install the camshaft

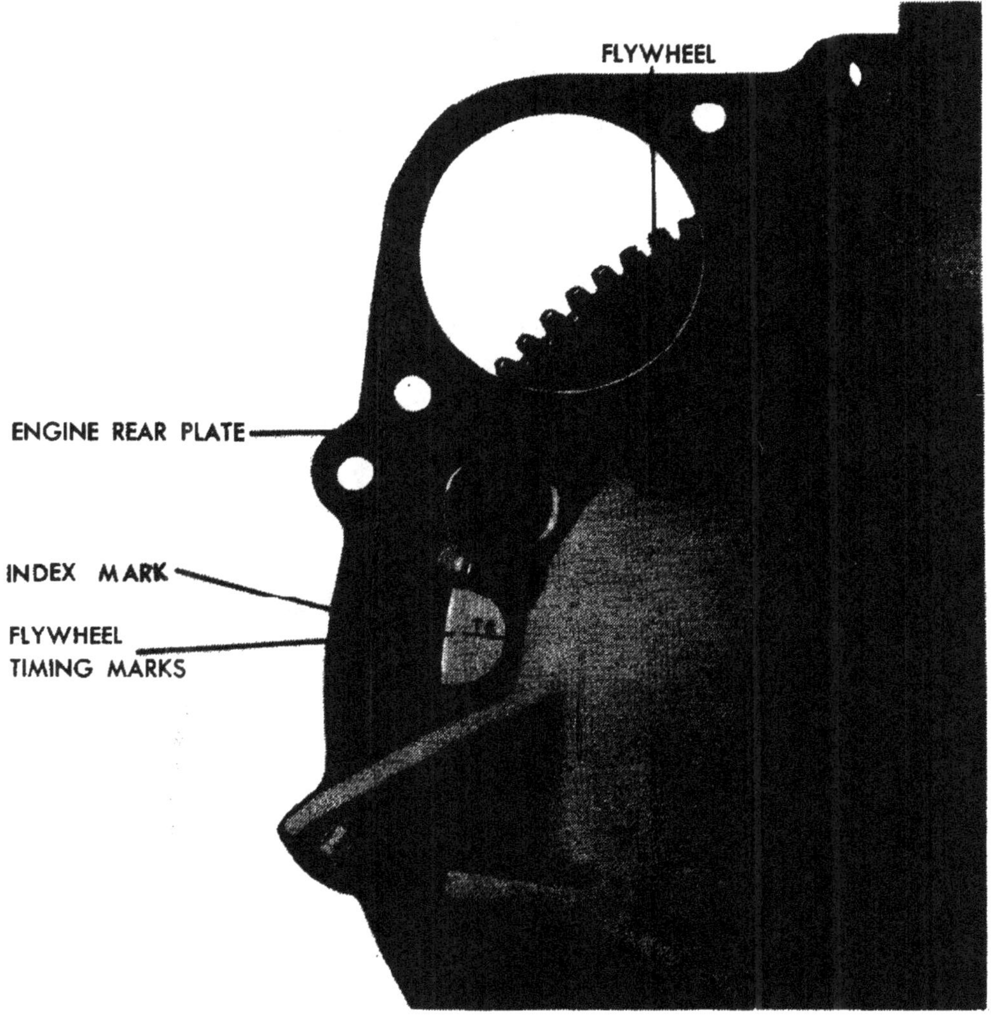

Figure 38 — Flywheel Timing Marks, T.C. (Top Center)

sprocket on the camshaft temporarily with two cap screws. Turn the camshaft sprocket until the punch mark on the camshaft sprocket is opposite the punch mark on the crankshaft sprocket (fig. 40). Remove the camshaft sprocket from the camshaft, being careful not to move the camshaft. Place the camshaft thrust washer on the camshaft. Place the camshaft drive chain on the crankshaft sprocket and camshaft sprocket, and install the camshaft sprocket on the camshaft with four lock washers and cap screws. Tighten the four cap screws, and bend the lock washer tabs down on the cap screws.

j. **Install Front Engine Cover.** Place the camshaft thrust plunger spring and plunger in the camshaft (fig. 25). Place a gasket on the engine front cover, and also install an oil seal in the recess provided in the cover. Install the cover on the engine.

Figure 39 — Installing Clutch Disk and Pressure Plate on Flywheel

k. Install Oil Pan. Hold a gasket and the oil intake float in place (fig. 11), and install the two lock washers and cap screws. Coat the bottom (machined surface) of the crankcase with grease, and install the oil pan gasket. Hold the oil pan in place, and install all the lock washers and cap screws except the six front cap screws. Hold the generator and fan drive pulley guard in place, and install the remaining six lock washers and cap screws. Tighten all the oil pan cap screws.

l. Install Cylinder Head. Install a cylinder head gasket on the cylinder block. Making sure there is no foreign material in the cylinders, place the cylinder head on the cylinder block, and install and tighten the cylinder head bolts to from 65 to 75 foot-pounds with a torque wrench. (Start with a centrally located bolt, and work alternately each way.)

m. Install Intake and Exhaust Manifold. Place an intake and exhaust manifold gasket in place on the cylinder block. Install the intake and exhaust manifold on the cylinder block. Install the seven

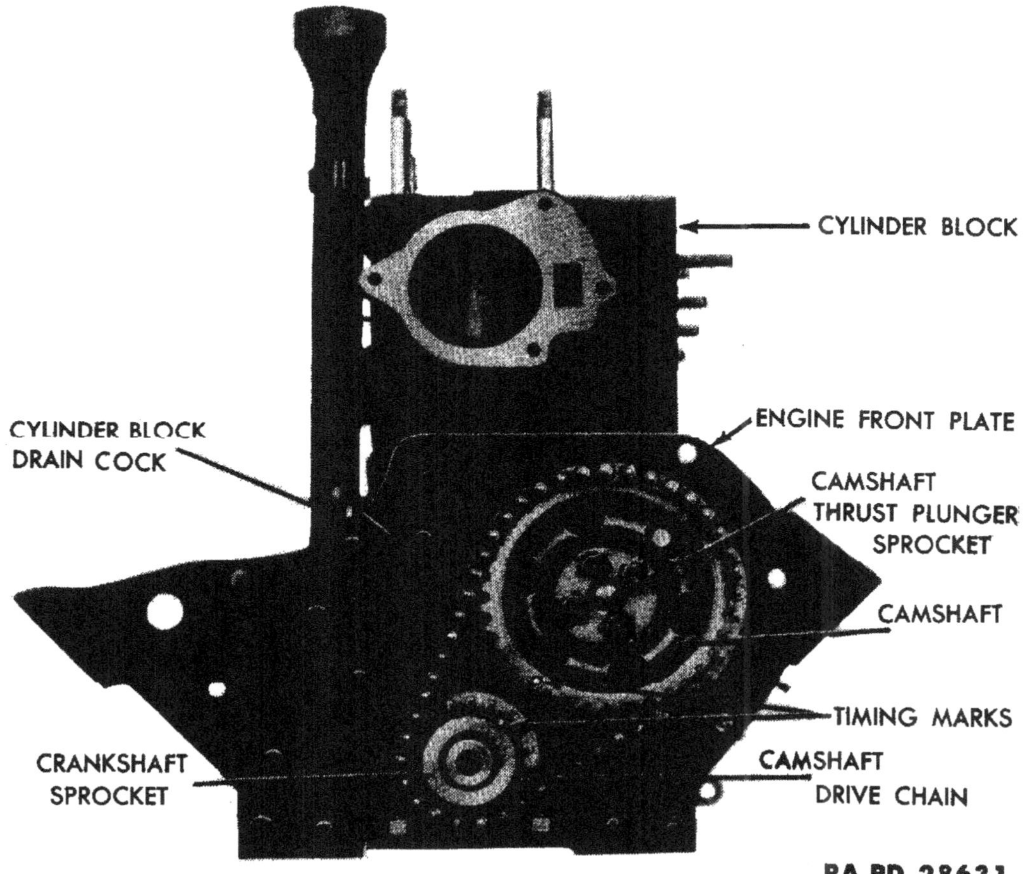

Figure 40 — Camshaft Timing Marks

flat washers and nuts. Connect the crankcase ventilation tube at the intake manifold, and at the crankcase ventilator assembly (fig. 41).

n. **Install Oil Pump.** Place a finger in No. 1 spark plug hole, and turn the crankshaft until No. 1 piston is coming up on compression stroke. Continue turning the crankshaft until the timing mark "IGN" appears on the flywheel and is in line with the index mark in the center of the timing hole on the engine rear plate (fig. 42). Install the distributor in the cylinder block (par. 19 e) temporarily. Set the rotor on No. 1 firing position (fig. 43) with the ignition points just breaking. Immerse the oil pump in a container of oil (same grade as used in engine), and turn the oil pump shaft assembly until the oil flows from the outlet hole in the oil pump body. Place a gasket on the oil pump and, with the wide side of pump shaft up, install the oil pump on the engine, making sure the slot in the oil pump shaft engages with the distributor shaft while the rotor is on No. 1 firing position with ignition

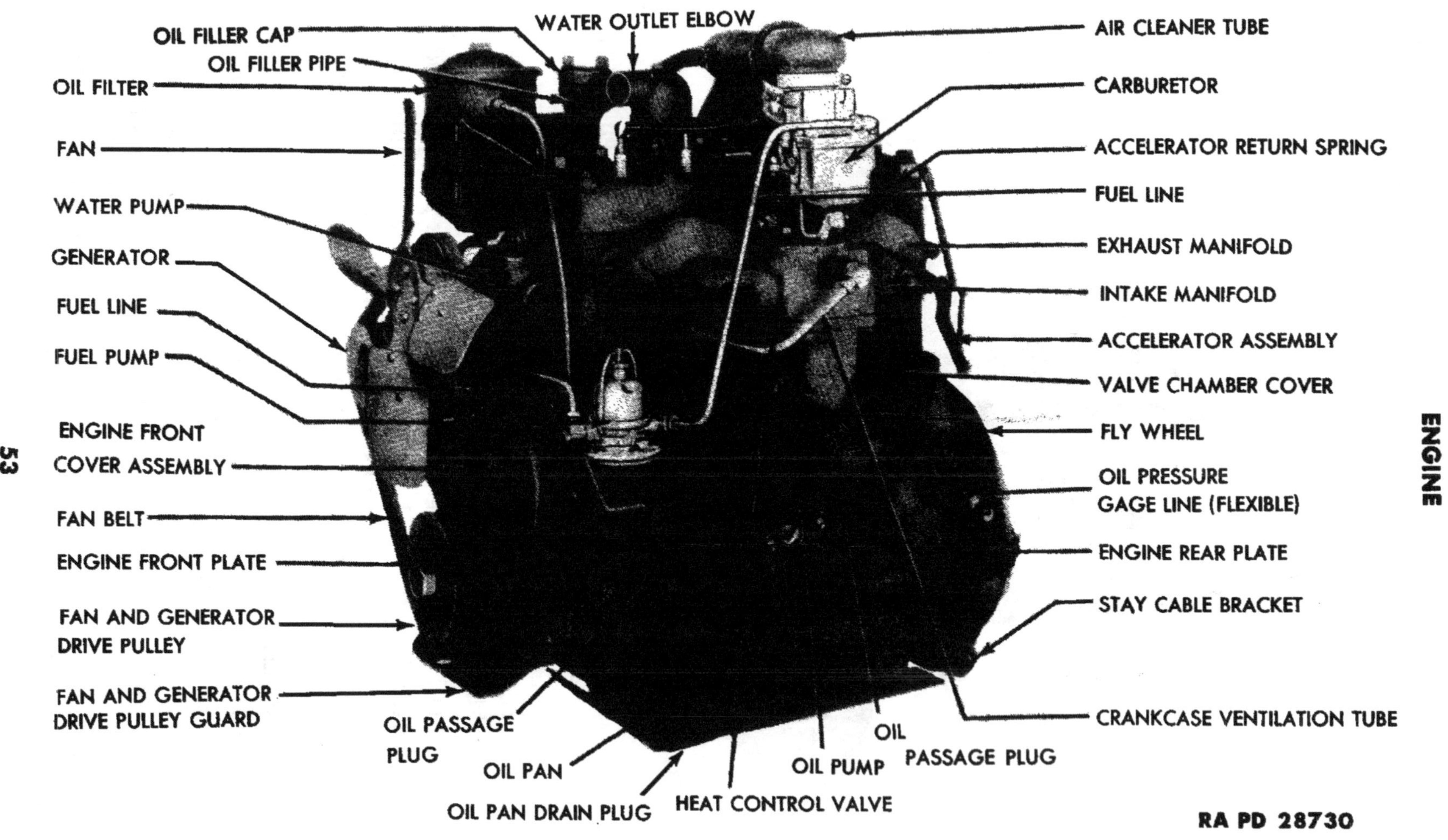

Figure 41 — Left Front View of Engine

Figure 42 — IGN Timing Marks on Flywheel

points just breaking. Install the lock washers and nuts on the oil pump. Remove the distributor.

o. **Install Water Outlet Elbow and Thermostat.** Install the thermostat and retainer in the water outlet elbow with the bellows of the thermostat facing downward. Place a gasket on the cylinder head and install the water outlet elbow, lock washers, and cap screws.

19. INSTALLATION OF ACCESSORIES.

a. **Install Water Pump.** Hold a gasket and the water pump in place on the engine, and install the three lock washers and cap screws.

b. **Install Carburetor.** Place a carburetor gasket and diffuser on the intake manifold and install the carburetor, accelerator return spring clip, lock washers, and nuts (fig. 41).

c. **Install Oil Filter** (fig. 43). Hold the oil filter in place, and install and tighten the three head nuts with a torque wrench to from 60 to 65 foot-pounds pull. Connect the oil filler pipe bracket to the oil filter bracket with a cap screw. Connect the outlet oil line to the engine front cover, and the inlet line to the elbow fitting located on the left-hand side of the engine in front of the fuel pump opening.

d. **Install Fuel Pump.** Place a gasket on the fuel pump. Hold the fuel pump in place on the engine, making sure the fuel pump rocker arm is on top of the camshaft. Install the two lock washers and cap screws in the fuel pump. Install the fuel line that connects the fuel pump and carburetor. Connect the generator brace and fuel line to the engine front plate. Connect the fuel line to the fuel pump.

e. **Install Distributor.** Place a thumb over No. 1 spark plug hole, and turn the crankshaft until No. 1 piston is coming up on compression stroke, and the timing mark "IGN" on the flywheel is in line with the index mark in the center of the timing hole on the engine rear plate (fig. 42). Install the distributor in the engine, and rotate the rotor until the distributor shaft engages in the oil pump shaft. Install the cap screw in the distributor hold-down clamp. Loosen the bolt in the distributor hold-down clamp, and turn the distributor until the points are just breaking. Tighten the bolt in the distributor hold-down clamp.

f. **Install Ignition Coil.** Install the ignition coil on the engine, making sure the bond strap is in place behind the ignition coil bracket (fig. 43). Connect the primary wire to the coil and distributor.

g. **Install Fan.** Hold the fan in place on the water pump pulley, and install the four lock washers and cap screws.

h. **Generator.** Install the generator bracket on the cylinder block with two lock washers and cap screws. Install the generator on the engine with the generator bolts, making sure there is a flat washer on each side of the rubber bushing in the generator bracket and engine front plate. Install a flat washer, lock washer, and nut on the gener. or front mount. Install a flat washer, bond strap, lock washer, and nut on the generator rear mount.

i. **Install Spark Plugs, Wires, and Air Cleaner Tubing.** Install air cleaner tube and bracket assembly, and tighten the head nuts from 60 to 65 foot-pounds. Connect the air cleaner tubing at the oil filler pipe (fig. 43) and carburetor. Install the distributor cap on the distributor. Pull the spark plug wires through the air cleaner tube bracket (fig. 43). Set the spark plug gap at 0.030 inch, and install the spark plugs and new spark plug gaskets in the cylinder head. Connect the spark plug wires.

Figure 43 — Right Side View of Engine

j. Install Accelerator Linkage. Install the accelerator (throttle linkage) on the engine with two lock washers and nuts. Connect the accelerator linkage to the carburetor throttle lever with a cotter pin. Connect the accelerator return spring (fig. 41) to the accelerator return spring clip.

k. Install Engine Front Supports. Install an engine front support on each side of the engine front plate.

Section VI

INSTALLATION OF ENGINE

Paragraph

20. INSTALLATION.

a. General. If the clutch housing was not removed with the engine, start the installation procedure, beginning with subparagraph **b** below. If the clutch housing was removed with the engine, assemble the clutch housing onto the engine, and install the clutch housing bolts; then proceed with the installation as outlined, starting with subparagraph **b** below. Some variation exists in the location of the various bond straps used to eliminate radio interference on these vehicles. Disregard references to bond straps in the following instructions, if they are not present on the particular vehicle being worked on. If bond straps are found in locations other than those mentioned in the following instructions, they must be connected when installing the engine.

b. Place Engine in Vehicle. Install a suitable engine sling or a rope on the engine (fig. 7). Lift the engine into the vehicle with a hoist, and lower the engine until the clutch disk in the engine is in line with the main drive gear shaft in the transmission. Place the gearshift lever in low speed position, and roll the vehicle forward and backward, at the same time pushing in on the engine until the splines on the main drive gear shaft are engaged with the splines in the clutch disk. Push the engine back onto the clutch housing, and install the clutch housing bolts. Lower the engine until the two hold-down bolts and bond strap can be installed loosely in the engine front support insulator. Lower the engine the rest of the way, and remove the engine sling or rope. Slide the stay cable through the bracket on the engine rear plate (fig. 4), and through the front crossmember. Install a nut on the stay cable, tighten it until all slack is removed, and install and tighten the stay cable lock nut. Tighten the two hold-down bolts in each engine front support insulator. Tighten the nut on each engine front support insulator.

c. **Install Clutch Control Lever and Cable.** Install the four lock washers and cap screws that hold the clutch housing to the transmission. Working through the inspection opening on top of the clutch housing, install the clutch control lever on the clutch release bearing carrier, and on the ball joint located on the main drive gear bearing retainer (fig. 44). Slide the clutch control lever cable through the hole in the clutch housing, and connect the clutch control lever cable yoke end to the clutch control lever and tube assembly with a clevis pin and cotter pin. Press the ball and socket joint end of the clutch control lever inward, and slide the clutch control lever cable in place on the clutch control lever (fig. 44).

d. **Install Cranking Motor.** Hold the cranking motor in place on the clutch housing, and install and tighten the two cranking motor cap screws on the clutch housing. Hold the cranking motor support bracket in place on the engine and install a lock washer, flat washer, generator bond strap, and cap screw; do not tighten the cap screw. Install and tighten a lock washer, flat washer, and cap screw in the cranking motor and cranking motor support bracket. Tighten the cap screw in the cranking motor support bracket. Connect the cranking motor cable to the cranking motor.

e. **Connect Oil Pressure and Water Temperature Gages.** Connect the water temperature gage (engine unit) at the right-hand side of the cylinder head (fig. 5). Connect the oil pressure gage line at the flexible oil line on the left-hand side of the engine.

f. **Connect Electrical Wires and Bond Straps.** Connect the field wire on the small post of the generator, and the armature wire and condenser on the large post. Connect the ground wire at the rear of the generator at the fillister-head cap screw. Connect the bond strap at the rear of the cylinder head. Connect the primary wire to the ignition coil.

g. **Connect Choke and Throttle Controls.** Slide the choke control cable and conduit through the choke lever carburetor bracket assembly on the carburetor, and the choke control cable through the collar on the choke lever. Push in the choke control button on the instrument panel. Pull the choke lever forward as far as possible, and tighten the set screw in the collar. Connect the throttle control cable and conduit to the choke lever carburetor bracket assembly with the carburetor air cleaner clamp, nut, and bolt. Run the throttle control cable and conduit to the left of the carburetor choker link between the link and the carburetor. Run the throttle control cable

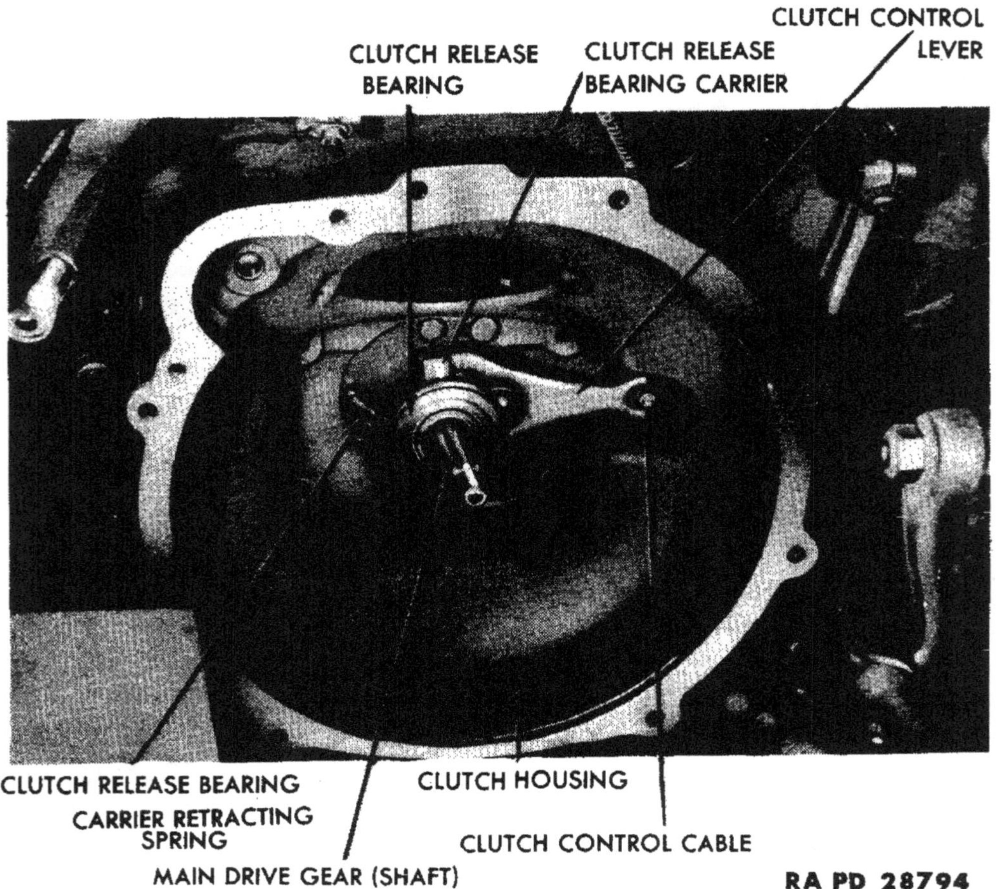

Figure 44 — Clutch Housing Installed in Vehicle

through the carburetor throttle shaft arm and screw assembly. Push the throttle control cable on the instrument panel all the way in. Tighten the screw on the carburetor throttle shaft arm.

h. Install Air Cleaner Hose. Slide the crankcase ventilation flexible hose on the oil filler pipe, and tighten the hose clamp. Slide the air cleaner flexible hose onto the metal tube, and tighten the clamp.

i. Connect Exhaust Pipe. Place a new exhaust gasket on the exhaust manifold, and attach the pipe to the manifold with a cap screw, lock washer, and bolt.

j. Install Radiator. Install the two carriage head bolts in the bottom of the radiator. Place a radiator pad on each of the two radiator mounting brackets. Lower the radiator into place in the vehicle. Install and tighten a flat washer and nut on each radiator bolt. Place a radiator bond strap on each radiator bolt, and install a flat washer and nut. Slide the radiator hose in place on the engine and radiator, and tighten the radiator hose clamps. Slide the straight end of the radiator brace through the radiator bracket mounted on top of the radiator; press the other end through the hole in the cowl (fig. 5), and install the lock washers and nuts.

k. Install Battery. Set the battery in the battery tray with the negative post toward the front of the vehicle. Install the battery hold-down frame, wing nuts, and battery cables.

l. Final Operations. Make sure the radiator and engine drain cocks are closed, and install the specified coolant. Tighten the oil pan drain plug, and install the specified amount and grade of oil. Start the engine. If the oil pressure does not register immediately on the oil pressure gage, stop the engine. Remove the oil pump relief valve retainer and prime the oil pump. Make adjustments and tests as outlined in paragraph 21.

21. ADJUSTMENTS AND TESTS IN VEHICLE.

a. Adjust Clutch. The free travel of the clutch pedal must be adjusted so that the pedal will have ¾-inch free travel before the clutch starts to disengage. Loosen the clutch control lever cable adjusting yoke lock nut. Turn the clutch control lever cable counterclockwise to decrease the pedal free travel, and clockwise to increase the pedal free travel. When a ¾-inch free pedal travel is obtained, tighten the clutch control lever cable adjusting yoke lock nut.

b. Set Timing With Neon Light.

(1) PRELIMINARY WORK. Loosen the screw in the timing hole cover, and move the cover to one side. Make certain the ignition switch is in the "OFF" position, then turn the engine with a crank until the timing mark "IGN" (fig. 42) on the flywheel is in line with the index mark on the engine rear plate. Mark the line under the timing mark "IGN" with chalk or white paint.

(2) CONNECT TIMING LIGHT (fig. 45). Attach the high tension lead of the timing light to the terminal of No. 1 spark plug. Attach the positive low tension lead to the positive terminal of the battery, and connect the negative low tension lead to the negative battery terminal.

(3) USE THE TIMING LIGHT. Start the engine, and allow it to warm up. Set the engine idle speed at 600 revolutions per minute. Point the timing light at the timing mark opening so that it can flash on the flywheel. If the timing mark "IGN" on the flywheel appears at the index mark on the opening in the engine rear plate, the timing is correct. If the timing mark "IGN" on the flywheel appears lower than the index mark, the timing is too far advanced.

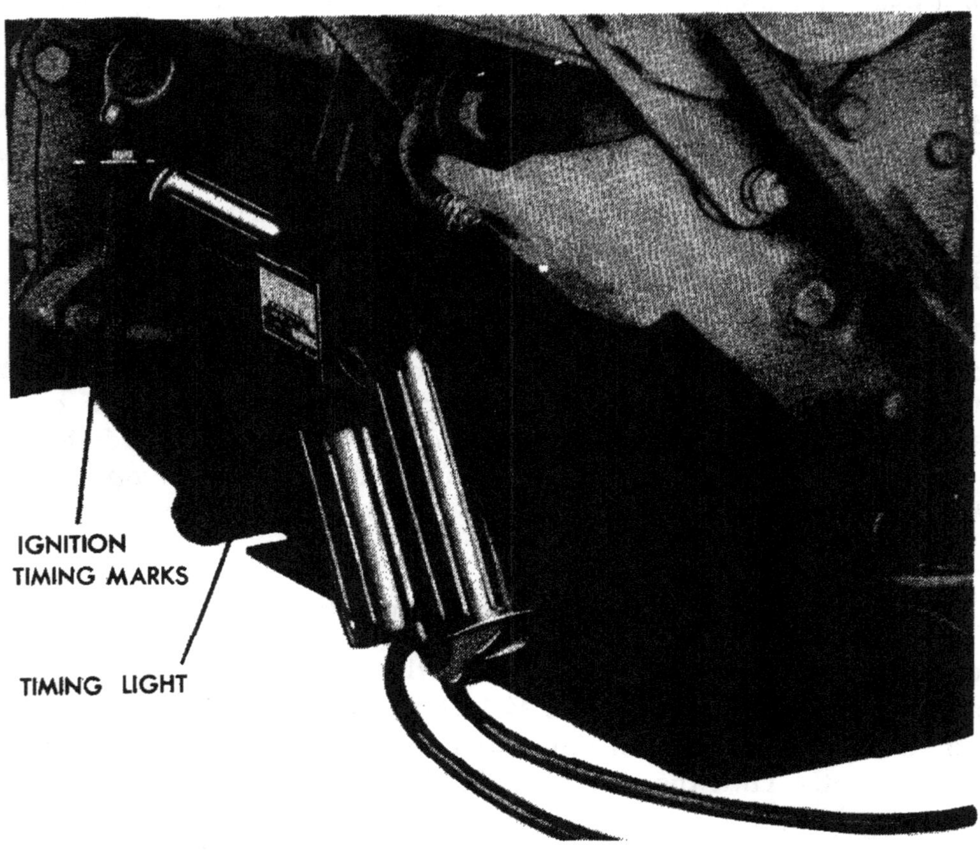

Figure 45 — Timing Engine, Using Timing Light (41-L-1440)

(4) ADJUST DISTRIBUTOR TO CORRECT TIMING. Loosen the bolt in the advance control arm, and turn the distributor clockwise to advance the ignition timing. Turn the distributor counterclockwise to retard the ignition timing. When the correct timing is obtained, tighten the bolt in the advance control arm. If the correct timing cannot be obtained by turning the distributor, set the timing as outlined in paragraph 19 e, but do not remove the distributor.

c. **Adjust Carburetor.** Start the engine, and allow it to run until it reaches normal operating temperature. Turn the idle fuel adjustment screw (fig. 1) clockwise, or counterclockwise, until all indications of vibration and roll are eliminated from the engine. Set the idle speed adjustment screw so that engine will idle at 400 revolutions per minute.

Section VII

FITS AND TOLERANCES

22. DEFINITION OF FITS.

a. **General.** The table of fits and tolerances (par. 23) gives the original clearance established between various parts at the time of manufacture, as well as wear and limit clearances that indicate to what point the clearance may increase before the parts must be replaced. These clearances all are based on the assumption that the parts involved are both at a temperature of 70° F. The following definitions of the various types of fits are given to assist in arriving at the correct amount of clearance required between parts not included in paragraph 23, as well as to give a better appreciation of the necessity for adhering to the various tolerances. Generally speaking, all bores are made to a standard size (so that standard reamers, plug gages, etc. may be used) with a plus tolerance. The maximum size of male parts is usually a standard size, less the minimum clearance required for the type of fit desired. The minimum size for male parts is the maximum size, minus the tolerance.

b. **Ring Fit.** A ring fit is the type of fit obtained when the two parts are of identical size. This is the type of fit required between a bore and a plug gage when using the plug gage to determine the inside diameter of the bore. With a ring fit, it is necessary to turn or ring the plug gage or part to force it through the bore. This type of fit does not provide space for film of oil.

c. **Slip Fit.** A slip fit exists when the male part is slightly smaller than the female part, and involves less clearance than a running fit (subpar. d below). An example of the minimum allowable clearance for a slip fit would be a piston pin that from its own weight would pass slowly through the connecting rod bushing (bushing and pin both in a vertical position). In most cases (except where only a limited movement of the parts is involved) slip fits are specified, when, due to anticipated expansion (subpar. g below) of the female part, enough additional clearance will result to change this type of fit to a running fit (subpar. d below), and provide adequate clearance for a film of oil.

d. **Running Fit.** A running fit is a fit providing enough clearance for a continuous film of oil between the two parts. A running fit usually requires 0.001 inch for the oil film plus a minimum of 0.001 inch clearance for each inch of diameter (subpar. g below).

e. **Press Fit.** A press fit is one that requires force to enter the male part into the bore. Accepted practice for press fits is to have the male part larger by 0.001 inch for each inch of diameter, than the bore into which it is to be pressed.

f. **Shrink Fit.** Generally speaking, a shrink fit is tighter than a press fit. The amount of the shrink ranges from 0.001 inch to 0.002 inch for each inch of diameter, and in some cases even more. There are two methods of shrinking two parts together, either one of which may be used (both may be used in some instances). One method involves expansion of the female member by heating. The second method involves contracting the male member by chilling with dry ice or liquid air.

g. **Effect of Expansion on Fits.** Allowances are made in establishing fits on parts that are exposed to high temperatures in order to provide for the anticipated expansion of the part during operation, and still provide adequate clearance for the type of fit required. Absolute minimum allowances for expansion of parts exposed to flame or exhaust gases (pistons, piston rings, and valves) is 0.001 inch for each inch of diameter or length. In anticipating the expansion of a valve stem or piston ring to make allowances for the additional gap required between the end of the valve and the push rod, or between the ends of the piston ring, 0.001 inch for each linear inch of the part is added.

23. FITS AND TOLERANCES.

CYLINDER BLOCK

Fit Location Name	Manufacturers Fit Tolerance	Fit Wear Limit	Type of Fit
Cylinder bore out-of-round	———	0.005 in.	———
Cylinder bore taper	———	0.010 in.	———
Clearance between camshaft and (front) bearing	0.002 in. to 0.0035 in.	0.006 in.	Running
Clearance between camshaft and intermediate and rear bearings	0.002 in. to 0.0035 in.	0.008 in.	Running
Valve guide and cylinder block	———	———	Press
Clearance between valve stem and valve guide (exhaust)	0.002 in. to 0.004 in.	0.005 in.	———
(intake)	0.0015 in. to 0.0035 in.	0.0045 in	———
Clearance between valve tappet and valve tappet bore	0.0005 in. to 0.002 in.	0.003 in.	———

CONNECTING ROD AND PISTON ASSEMBLY

Fit Location Name	Manufacturers Fit Tolerance	Fit Wear Limit	Type of Fit
Connecting rod side clearance	0.005 in. to 0.009 in.	0.013 in.	———
Connecting rod clearance on crankshaft	0.0005 in. to 0.001 in.	0.001 in.	Running
Piston pin clearance in connecting rod.	Locked in connecting rod	———	———
Piston pin clearance in piston	0.0001 in. to 0.0005 in.	0.0015 in.	Slip
Piston and cylinder at skirt	5-pound to 10-pound pull with 0.003-in. feeler	5-pound to 10-pound pull with 0.010-in. feeler	———
Piston top land and cylinder	0.0205 in. to 0.0225 in.	———	———
Piston ring to groove clearance (all rings)	0.0005 in. to 0.0015 in.	0.003 in.	———
Piston ring gap (all rings)	0.008 in. to 0.013 in.	0.013 in.	———
VALVES AND VALVE SPRINGS			
Intake valve stem diameter	0.373 in.	0.368 in.	———
Valve seat angle	45 degree	———	———
Exhaust valve stem diameter	0.3725 in.	0.3685 in.	———
		———	———
Valve seat angle	45 degree		
Spring tension at 2 7/64 inches (all springs)	50 pounds	———	———
Spring tension at 1¾ inches (all springs)	116 pounds	———	———
OIL PUMP			
Clearance between pinion shaft and pinion gear	0.001 in. to 0.003 in.	0.003 in.	Running
Clearance between oil pump housing and oil pump shaft	0.002 in. to 0.004 in.	0.005 in.	Running
Oil pump relief plunger spring tension at 1 1/16 in.	5½ pounds	———	———
CRANKSHAFT			
Crankshaft end play	0.004 in. to 0.006 in.	0.006 in.	———
Main bearing clearance (all main bearings)	0.0005 in. to 0.001 in.	0.001 in.	Running

24. TORQUE WRENCH READINGS.

Main bearing nuts	65 to 70 ft-lb
Connecting rod nuts	50 to 55 ft-lb
Flywheel to crankshaft cap screws	36 to 40 ft-lb
Cylinder head nuts	60 to 65 ft-lb
Cylinder head bolts	65 to 75 ft-lb

CHAPTER 3

CLUTCH ASSEMBLY

25. DESCRIPTION AND DATA.

a. **Description.** The clutch is of the single-plate automotive type, composed of two major units (fig. 46); the pressure plate assembly, and the driven plate or disk. The pressure plate is adjusted at the factory, and requires no other adjustments, except where it is necessary to install new clutch pressure springs, clutch fingers, or pressure plate.

b. **Data.**

Type	Single dry plate
Torque capacity	132 ft-lb
Clutch disk:	
Make	Borg and Beck
Facings	1 woven, 1 molded asbestos
Facing diameter	Inside 5⅛ in. Outside 7⅞ in.
Facing thickness	0.125 in.
Pressure plate:	
Make	Atwood
Number of springs	3
Spring pressure at 1$\frac{9}{16}$ in.	220 to 230 lb

26. PRESSURE PLATE DISASSEMBLY.

a. **Remove Clutch Adjusting Screw** (fig. 47). Place the pressure plate in a press with a wood block 2½ inches square on top of the clutch fingers (fig. 48). Depress the clutch fingers, and remove the three clutch adjusting screws, lock nuts, and lock washers. Release the pressure on the clutch fingers slowly to prevent the clutch pressure springs from flying out from under the clutch fingers.

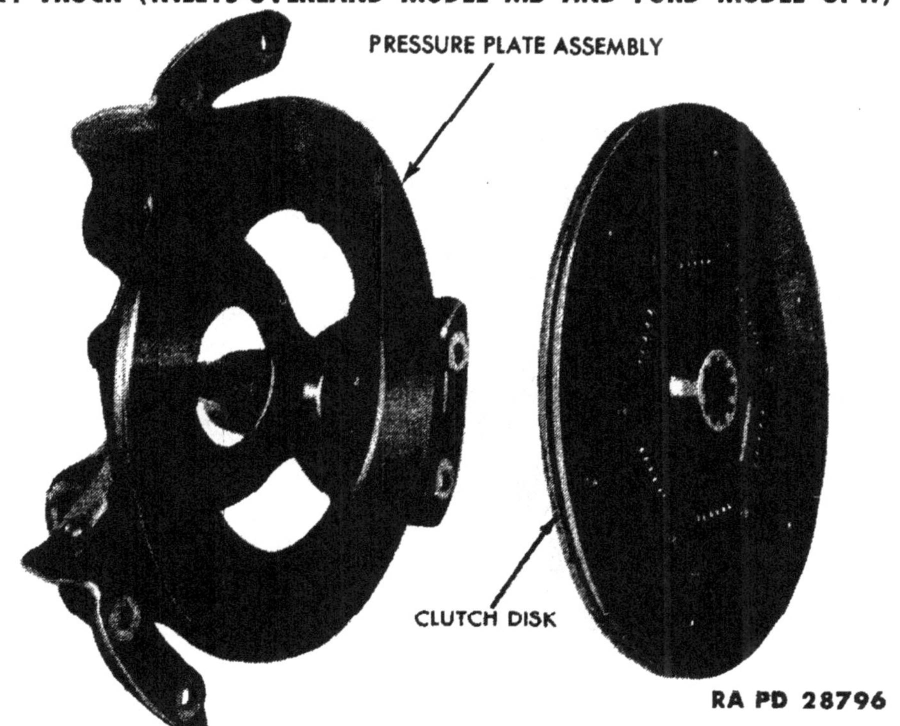

Figure 46 — Clutch Disk and Pressure Plate

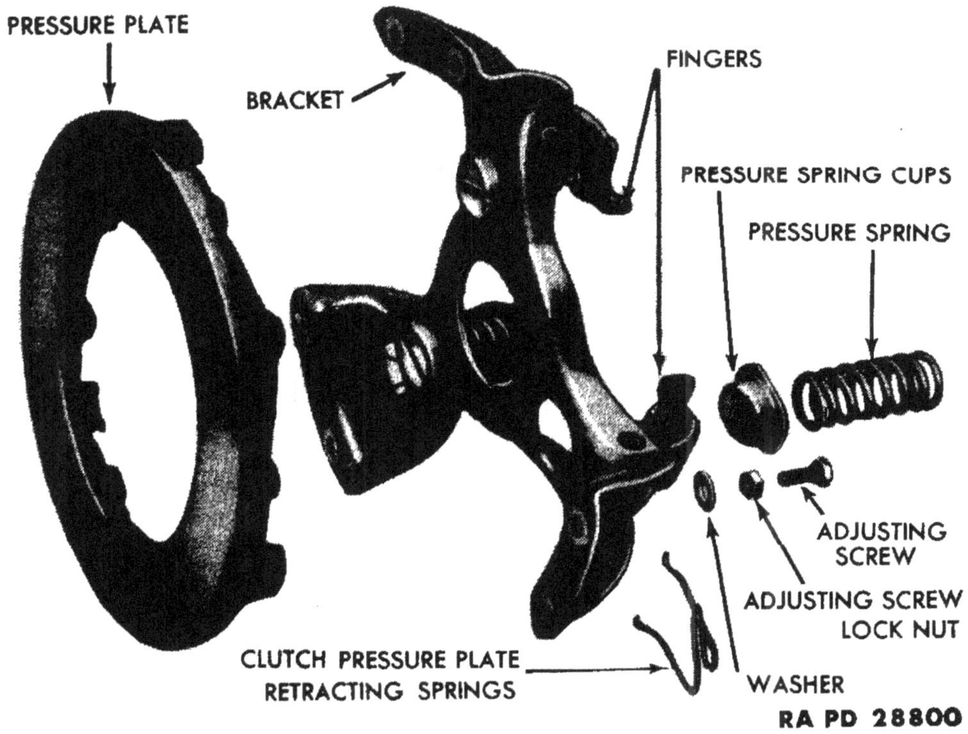

Figure 47 — Pressure Plate Disassembled

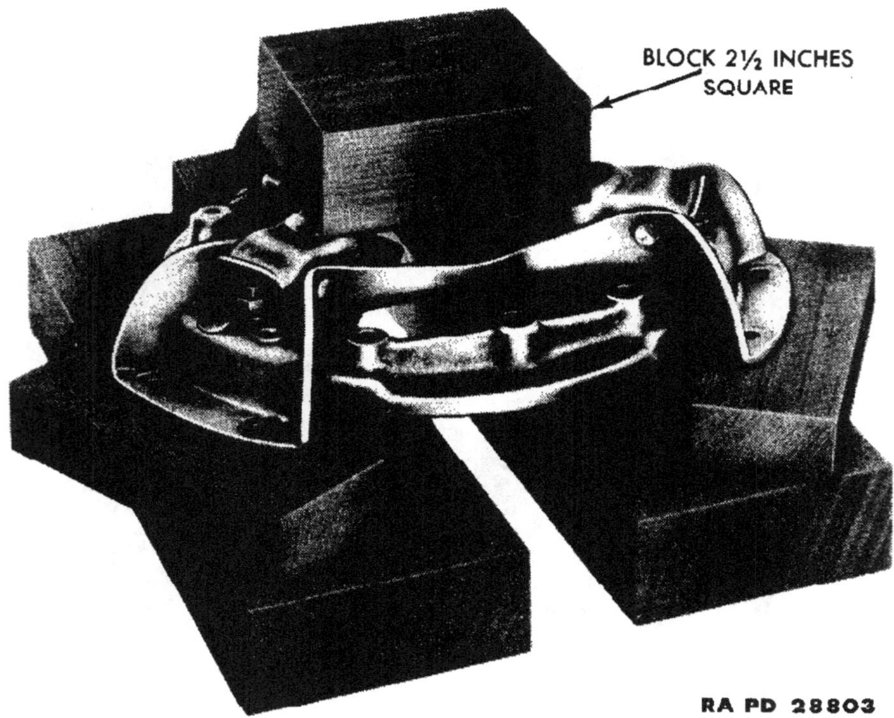

Figure 48 — Pressure Plate Blocked Up in Press for Disassembling or Assembling

b. Remove Clutch Pressure Spring Cups and Springs. Push the clutch pressure spring cups and clutch pressure springs from the pressure plate bracket with a screwdriver or punch (fig. 49). Remove the clutch pressure plate return springs from the clutch pressure plate bracket.

27. CLEANING, INSPECTION, AND REPAIR.

a. Cleaning. Clean all parts thoroughly in dry-cleaning solvent.

b. Inspection and Repair.

(1) CLUTCH PRESSURE PLATE (fig. 47). A ridged, scored, radial cracked, or burned pressure plate must be replaced.

(2) PRESSURE PLATE BRACKET (fig. 47). A distorted pressure plate bracket or a pressure plate bracket with worn clutch fingers must be replaced.

(3) CLUTCH PRESSURE SPRINGS. Place each clutch pressure spring in a tension scale, and depress it to 1$\frac{9}{10}$ inches (fig. 50). If

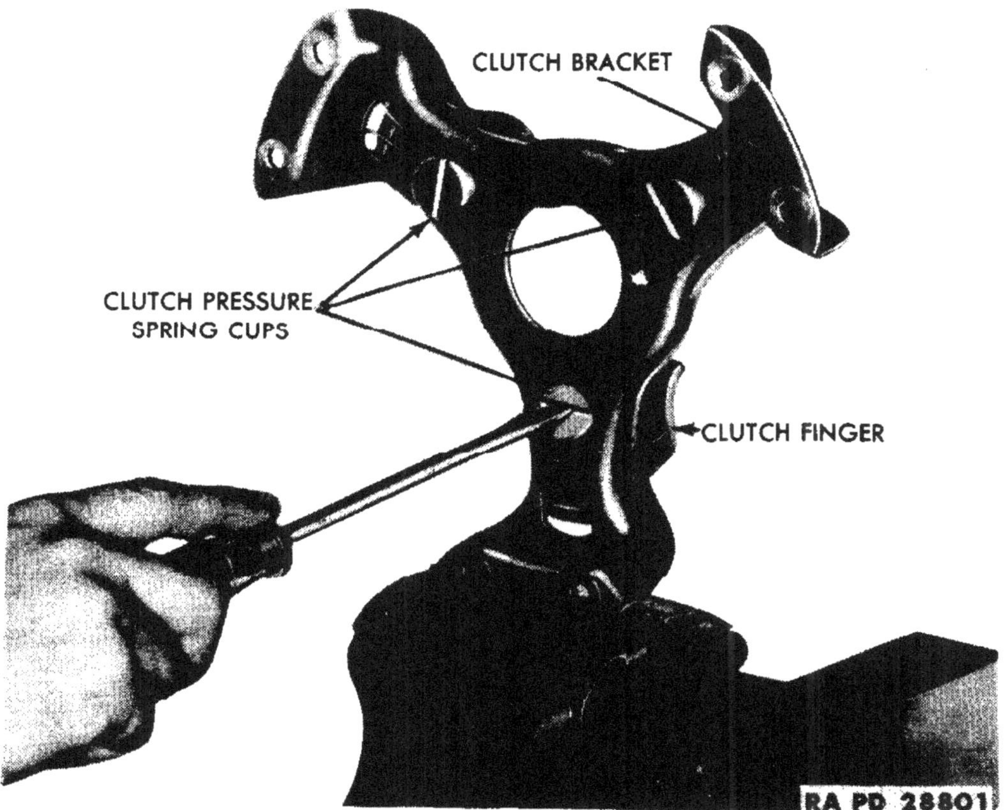

Figure 49 — Removing Clutch Pressure Spring Cups and Clutch Pressure Springs

the spring tension is less than 220 pounds on any clutch pressure spring, it must be replaced.

28. ASSEMBLY OF PRESSURE PLATE.

a. Install Clutch Pressure Spring Cups and Springs. Install the clutch pressure spring cups and clutch pressure springs in the pressure plate bracket (fig. 51), making sure the indentation of each clutch pressure spring cup is toward the center of the pressure plate bracket (fig. 51).

b. Install Pressure Plate on Pressure Plate Bracket. Slide a pressure plate return spring in place under each clutch finger on the pressure plate bracket (fig. 47). Place the pressure plate bracket in a press, blocking it up as shown in figure 48. Place a wood block 2½ inches square on top of the clutch fingers, and depress the fingers (fig. 48). Install the three clutch adjusting screws, lock nuts, and flat washers in the pressure plate.

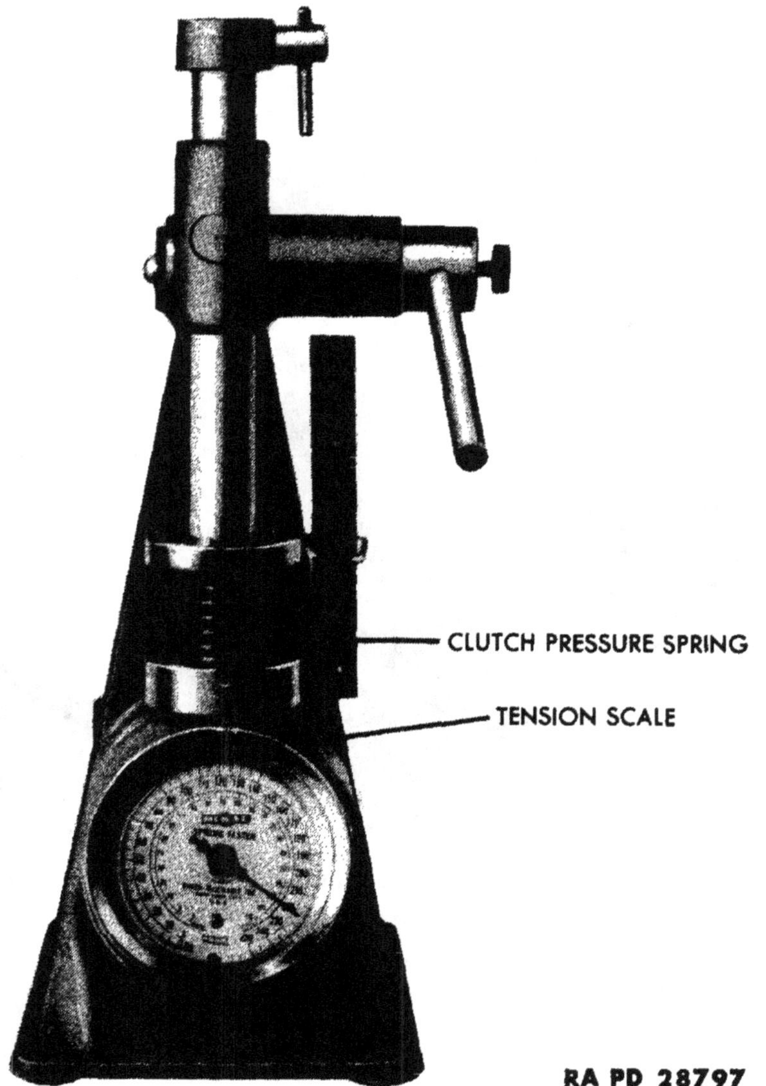

Figure 50 — Testing Spring Tension of Clutch Pressure Spring, Using Tester (41-T-1600)

c. **Adjust Pressure Plate.** Install the clutch disk and pressure plate onto the flywheel (par. 18 **h**). Hold a straightedge across the clutch fingers, and measure the distance from the straightedge to the face of the pressure plate bracket (fig. 52). Turn each clutch adjusting screw until a distance of $2\frac{7}{32}$ inches is established between the straightedge and the face of the pressure plate bracket. Hold the clutch adjusting screws with a wrench, and tighten each clutch adjusting screw lock nut. Recheck the distance between the straightedge and the face on the pressure plate bracket.

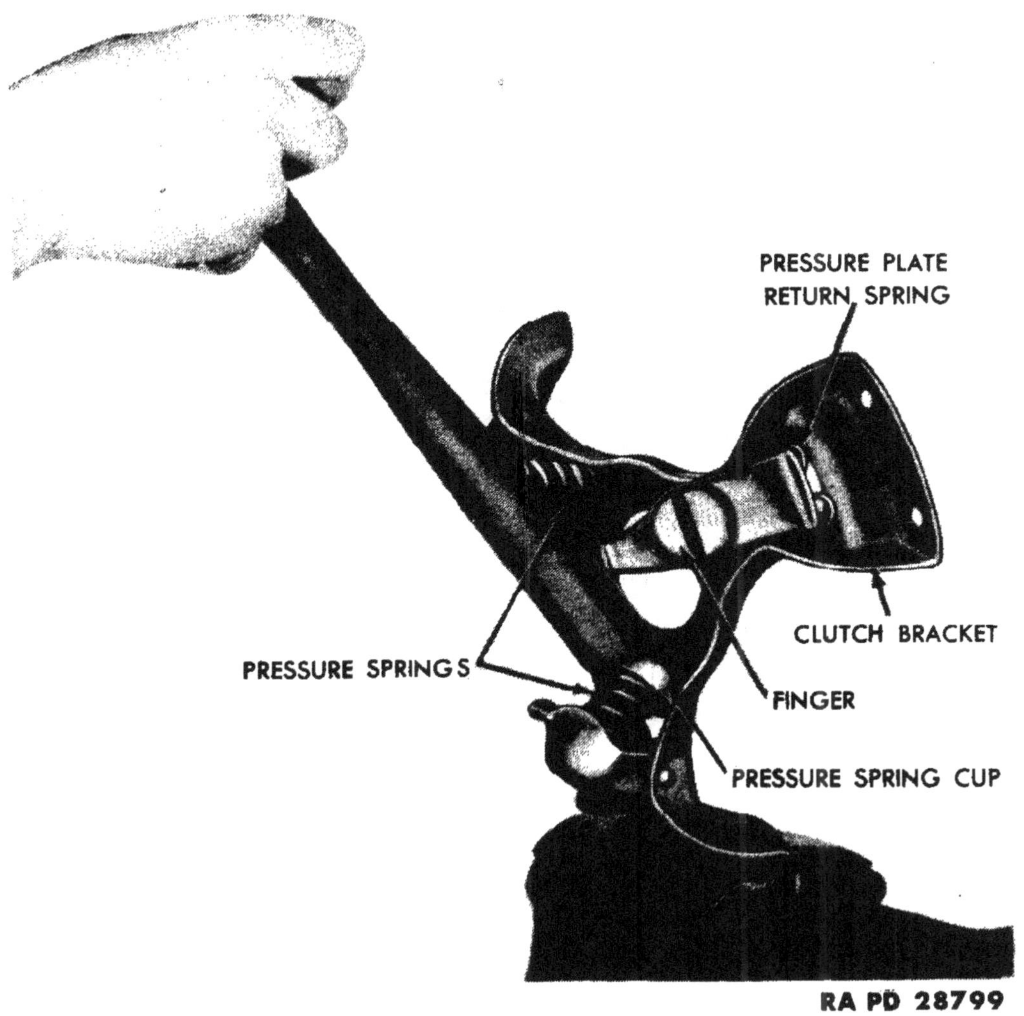

Figure 51 — Installing Clutch Pressure Spring Cups and Pressure Springs in Pressure Plate Bracket

CLUTCH ASSEMBLY

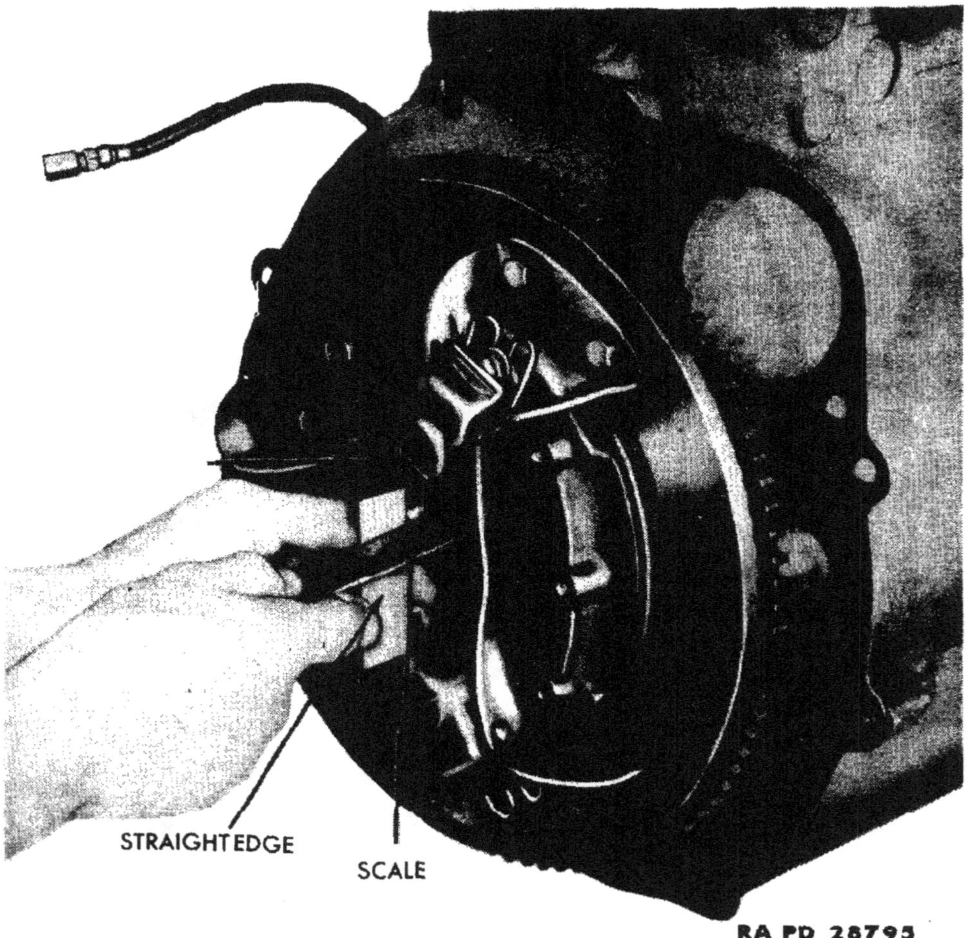

Figure 52 — Adjusting Pressure Plate

REFERENCES

PUBLICATIONS INDEXES.

The following publications indexes should be consulted frequently for latest changes to or revisions of the publications given in this list of references and for new publications relating to materiel covered in this manual:

Introduction to Ordnance Catalog (explaining SNL system)	ASF Cat. ORD1 IOC
Ordnance Publications for Supply Index (index to SNL's)	ASF Cat. ORD2 OPSI
Index to ordnance publications (listing FM's, TM's, TC's, and TB's of interest to ordnance personnel, MWO's, OPSR, BSD, S of SR's, OSSC's and OFSB's and includes Alphabetical List of Major Items with Publications Pertaining Thereto)	OFSB 1-1
List of Publications for Training (listing MR's, MTP's, T/BA's, T/A's, and FM's, TM's, and TR's concerning training)	FM 21-6
List of Training Films, Film Strips, and Film Bulletins (listing TF's, FS's, and FB's by serial number and subject)	FM-21-7
Military Training Aids (listing Graphic Training Aids, Models, Devices, and Displays)	FM 21-8

STANDARD NOMENCLATURE LISTS.

Cleaning, preserving and lubrication materials, recoil fluids, special oils, and miscellaneous related items	SNL K-1
Soldering, brazing and welding materials, gases and related items	SNL K-2
Tools, maintenance for repair of automotive vehicles	SNL G-27 Volume 1.

REFERENCES

Tool-sets, for ordnance service command automotive shops	SNL N-30
Tool-sets, motor transport	SNL N-19
Truck, ¼-ton, 4x4, command reconnaissance (Ford and Willys)	SNL G-503

EXPLANATORY PUBLICATIONS.

Fundamental Principles.

Automotive electricity	TM 10-580
Automotive lubrication	TM 10-540
Basic Maintenance Manual	TM 38-250
Driver's Manual	TM 10-460
Electrical fundamentals	TM 1-455
Military motor vehicles	AR 850-15
Motor vehicle inspections and preventive maintenance service	TM 9-2810
Precautions in handling gasoline	AR 850-20
Standard Military Motor Vehicles	TM 9-2800
The internal combustion engine	TM 10-570

Maintenance and Repair.

Cleaning, preserving, lubricating and welding materials and similar items issued by the Ordnance Department	TM 9-850
Cold weather lubrication and service of combat vehicles and automotive materiel	OFSB 6-11
Maintenance and care of pneumatic tires and rubber treads	TM 31-200
Ordnance Maintenance: Power train, chassis, and body for ¼-ton 4x4 truck (Ford and Willys)	TM 9-1803B
Ordnance Maintenance: Electrical equipment (Auto-Lite)	TM 9-1825B

Ordnance Maintenance: Hydraulic Brake System (Wagner) .... TM 9-1827C

Ordnance Maintenance: Carburetors (Carter) .... TM 9-1826A

Ordnance Maintenance: Fuel Pumps .... TM 9-1828A

Ordnance Maintenance: Speedometers and Tachometers (Stewart-Warner) .... TM 9-1829A

Tune-up and adjustment .... TM 10-530

Protection of Materiel.

Camouflage .... FM 5-20

Chemical decontamination, materials and equipment .... TM 3-220

Decontamination of armored force vehicles .... FM 17-59

Defense against chemical attack .... FM 21-40

Explosives and demolitions .... FM 5-25

Storage and Shipment.

Ordnance storage and shipment chart, group G—Major items .... OSSC-G

Registration of motor vehicles .... AR 850-10

Rules governing the loading of mechanized and motorized army equipment, also major caliber guns, for the United States Army and Navy, on open top equipment published by Operations and Maintenance Department of Association of American Railroads.

Storage of motor vehicle equipment .... AR 850-18

INDEX

INDEX

WAR DEPARTMENT TECHNICAL MANUAL

ORDNANCE MAINTENANCE

Power Train, Body, and Frame for 1/4-Ton 4x4 Truck

(Willys-Overland Model MB and Ford Model GPW)

WAR DEPARTMENT

WAR DEPARTMENT
Washington 25, D. C., 8 April 1944

TM 9-1803B, Ordnance Maintenance: Power Train, Body, and Frame for ¼-ton 4 x 4 Truck (Willys-Overland Model MB and Ford Model GPW), is published for the information and guidance of all concerned.

[A.G. 300.7 (17 Nov 43)
O.O.M. 461/(TM-9) Rar. Ars. (4-15-44)]

BY ORDER OF THE SECRETARY OF WAR:

G. C. MARSHALL,
Chief of Staff.

OFFICIAL:
J. A. ULIO,
Major General,
The Adjutant General.

DISTRIBUTION: R 9 (4); Bn 9 (2); C 9 (5).

(For explanation of symbols, see FM 21-6.)

CONTENTS

★This Technical Manual supersedes TB 1803-1, dated 8 December 1943. For supersession of Quartermaster Corps 10-series Technical Manuals, see paragraph 1 j.

CHAPTER 1

INTRODUCTION

1. SCOPE.

a. The instructions contained in this manual are for the information and guidance of personnel charged with the maintenance and repair of the power train, body, and frame of the ¼-ton 4x4 truck. These instructions are supplementary to field and technical manuals prepared for the using arms. This manual does not contain information which is intended primarily for the using arms, since such information is available to ordnance maintenance personnel in 100-series TM's or FM's.

b. This manual contains a description of, and procedure for, removal, disassembly, inspection, and repair of the transmission, transfer case, axles, body, and frame.

c. TM 9-803 contains operating instructions and information for the using arms.

d. TM 9-1803A contains instructions for the information and guidance of personnel charged with the maintenance and repair of the 4-cylinder engine used in these vehicles.

e. TM 9-1825B contains information for the maintenance of the Auto-Lite electrical equipment.

f. TM 9-1826A contains information for the maintenance of the Carter carburetor.

g. TM 9-1827C contains information for the maintenance of the Wagner hydraulic brake system.

h. TM 9-1828A contains information for the maintenance of the A. C. fuel pump.

i. TM 9-1829A contains information for the maintenance of the speedometer.

j. This manual includes pertinent ordnance maintenance instructions from the following Quartermaster Corps 10-series Technical Manuals. Together with TM 9-803 and TM 9-1803A, this manual supersedes them:

(1) TM 10-1103, dated 20 August 1941.

(2) TM 10-1207, dated 20 August 1941.

(3) TM 10-1349, dated 3 January 1942.

(4) TM 10-1513, Changes 1, dated 15 January 1943.

RA PD 28742

Figure 1 — ¼-ton Truck 4 x 4 — Three-quarter Front View

2. MWO AND MAJOR UNIT ASSEMBLY REPLACEMENT RECORD.

a. **Description.** Every vehicle is supplied with a copy of AGO Form No. 478 which provides a means of keeping a record of MWO's completed or major unit assemblies replaced. This form includes spaces for the vehicle name and U. S. A. Registration Number, instructions for use, and information pertinent to the work accomplished. It is very important that this form be used as directed and that it remain with the vehicle until the vehicle is removed from service.

b. **Instructions for Use.** Personnel performing modifications or major unit assembly replacements must record clearly on the form, a description of the work completed, and must initial the form in the columns provided. When each modification is completed, record the date, hours and/or mileage, and MWO number. When major unit assemblies, such as engine, transmission, transfer case, are replaced, record the date, hours and/or mileage and nomenclature of the unit assembly. Minor repairs and minor parts and accessory replacements need not be recorded.

c. **Early Modifications.** Upon receipt of a vehicle for modification or repair, by a third or fourth echelon repair facility, maintenance personnel will record the MWO numbers of modifications applied prior to the date of AGO Form No. 478.

CHAPTER 2

POWER TRAIN

Section I

POWER TRAIN DESCRIPTION

3. POWER TRAIN DESCRIPTION.

a. The power from the engine is transmitted to the driving wheels through a transmission and a transfer case, each of which provides a means of selecting the gear reduction. The power from the transfer case is transmitted to the front and rear axles through propeller shafts equipped with universal joints. The transmission is located at the rear of the engine and is secured to the clutch housing (fig. 2). The various gears in the transmission (par. 4) are controlled by a shift lever. The transfer case is mounted directly onto the rear of the transmission. The transmission output shaft extends from the rear of the transmission into splines of the main drive gear in the transfer case. The transfer case is provided with two levers, one to select the transfer case ratio, and the other to engage or disengage the front axle (fig. 5). A hand brake drum is mounted on the rear axle output shaft. Each axle is of the spiral bevel hypoid gear full-floating type, equipped with the conventional differential.

Section II

TRANSMISSION

4. DESCRIPTION AND DATA.

a. **Description.** The transmission (fig. 3) is of the 3-speed type with synchronized second and high speed gears. The transmission and transfer case are mounted on rubber on the frame center cross-member. The gearshift lever is incorporated in the gearshift housing.

b. **Data.**

Make .... Warner
Model T84J
Type .... Synchronous Mesh
Speeds:
  Forward 3
  Reverse 1
Ratios:
  Low .... 2.665 to 1
  Second .... 1.564 to 1

Figure 2 — Power Train

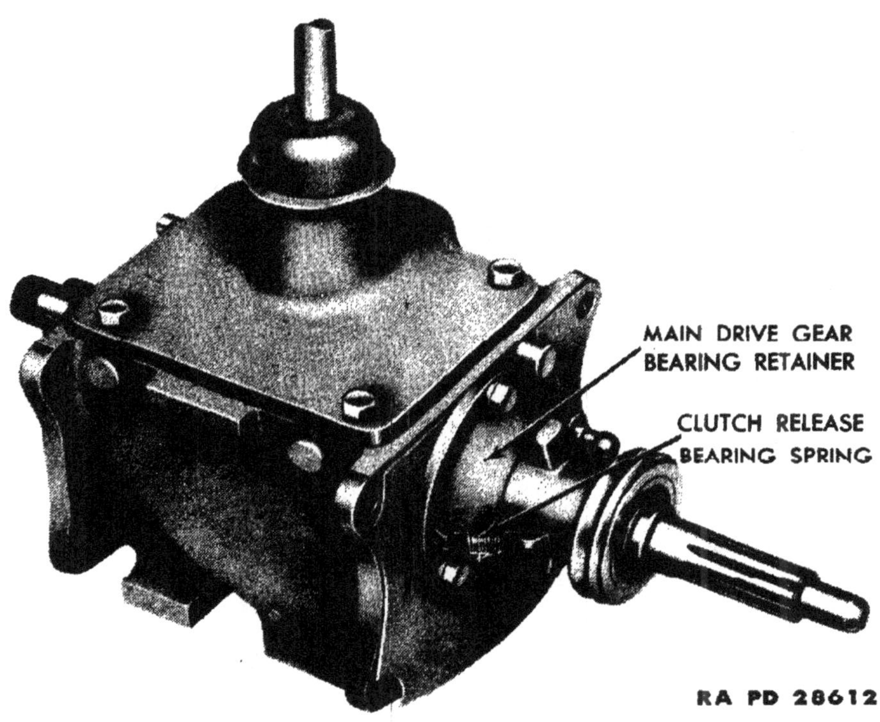

Figure 3 — Transmission — Three-quarter Front View

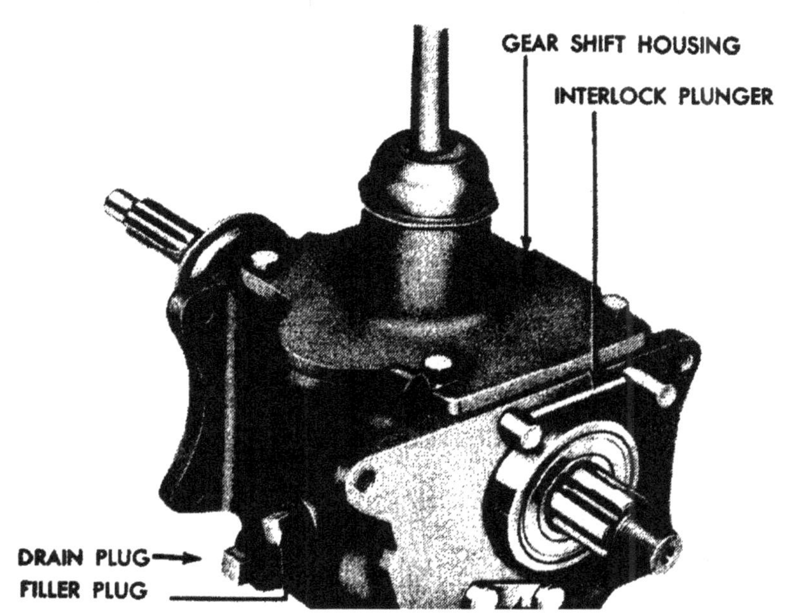

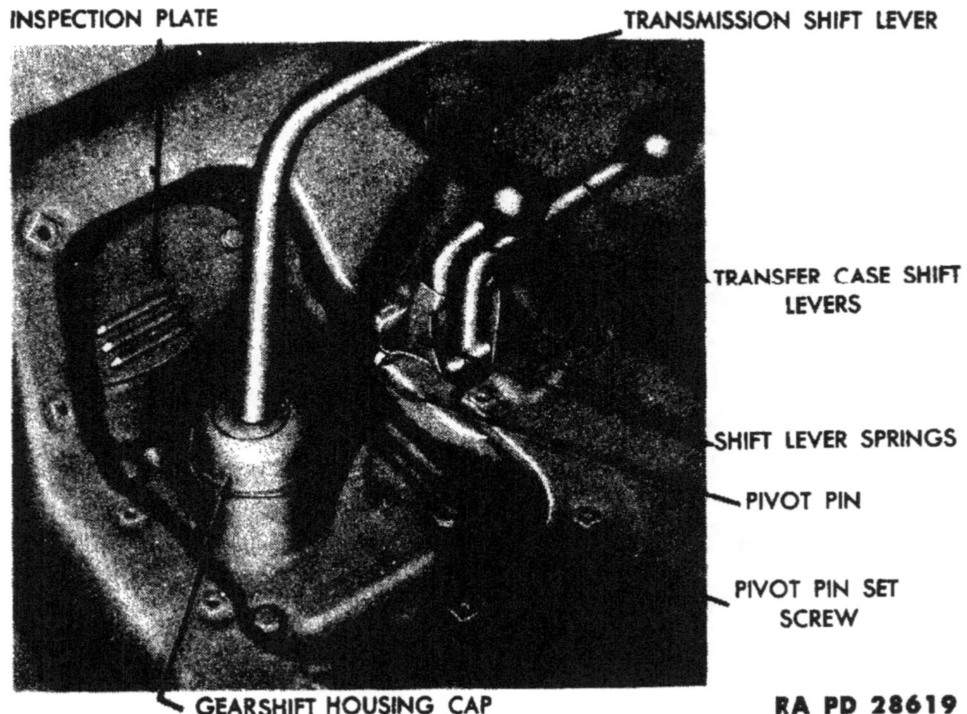

Figure 5 — Transmission and Transfer Case Shift Levers

High	1 to 1
Reverse	3.554 to 1

Bearings:

Clutch shaft (flywheel)	Bushing
Clutch release	Ball
Clutch shaft rear (main drive gear)	Ball
Mainshaft front	13 rollers
Mainshaft rear	Ball
Countershaft gear	Bushings (2)
Reverse idle gear	Bushing

5. REMOVAL.

a. **Remove Floor Plate and Shift Lever** (fig. 5). Remove the cap screws from the floor plate at the transmission, and remove the floor plate. Remove the gearshift housing cap and remove the shift lever from the transmission. Remove the set screw that secures the shift lever pivot pin on the transfer case and, with a suitable drift, remove the shift lever pivot pin. Remove the two shift levers and shift lever springs from the transfer case. Remove the two cap screws that secure the clutch housing inspection plate and remove the inspection plate.

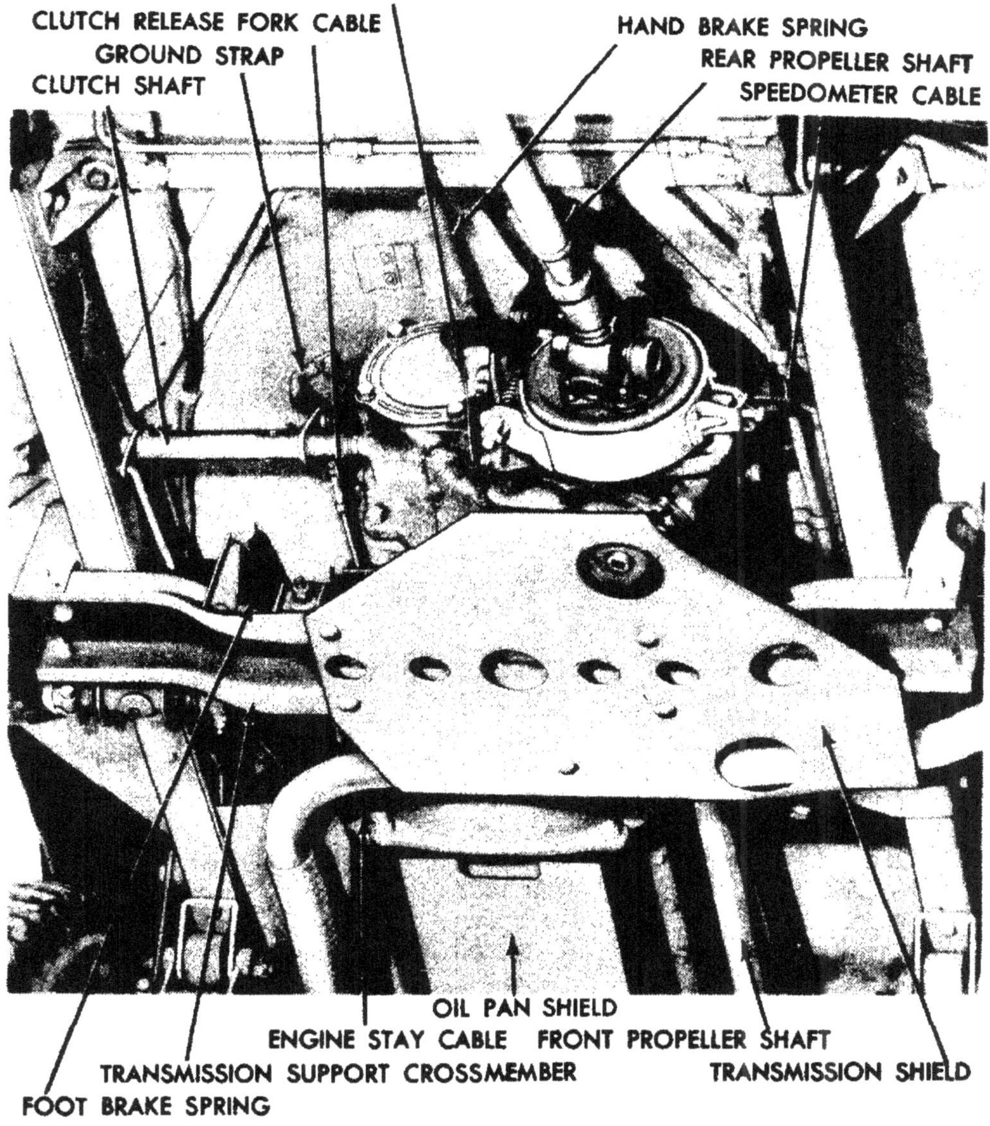

Figure 6 — Under Side of Chassis

b. **Remove Transmission Shield** (fig. 6). Remove the cap screws that secure the exhaust pipe clamp to the shield, and remove the clamp. Remove the five bolts that secure the transmission shield to the transmission support crossmember. Remove the transmission shield.

c. **Remove Brake Springs and Speedometer Cable** (fig. 6). Remove the hand brake spring. Remove the foot brake spring leading from the bottom of the brake pedal to the transmission support crossmember. Disconnect the speedometer cable at the transfer case.

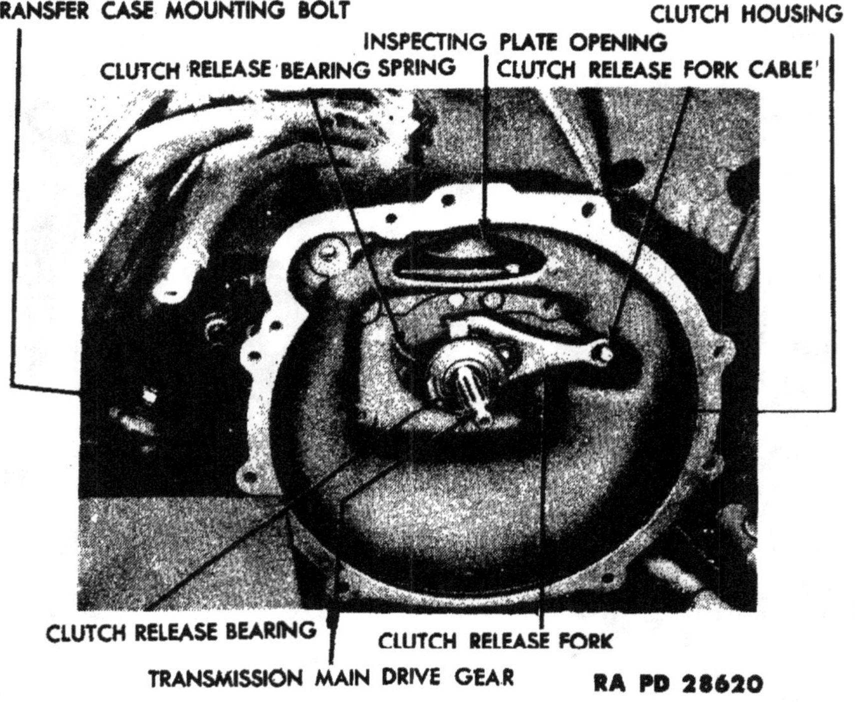

Figure 7 — Clutch Release Fork

d. **Remove Hand Brake Cable, Clutch Cable, and Engine Stay Cable** (fig. 6). Remove the clevis pin that secures the hand brake cable to the brake band. Remove the hand brake cable clamp at the transfer case. Disconnect the clutch cable at the clutch shaft. Remove the two nuts from the engine stay cable on the transmission support crossmember and remove the engine stay cable.

e. **Remove Propeller Shafts** (fig. 6). Disconnect the front propeller shaft at the transfer case (par. 17 a). Disconnect the rear propeller shaft at the transfer case (par. 17 b).

f. **Remove Ground Strap** (fig. 6). Remove the ground strap leading from the transfer case to the floor plate.

g. **Remove Clutch Release Fork** (fig. 7). Working through the inspection plate opening on the clutch housing, remove the clutch cable from the clutch release fork, and remove the clutch release fork from the clutch housing.

h. **Disconnect Radiator Hose.** Drain the coolant from the radiator. Loosen the radiator hose clamp at the radiator end, and remove the hose from the radiator.

i. **Disconnect Transmission at Clutch Housing** (fig. 6). Place a jack under the oil pan shield at the rear of the engine. Remove

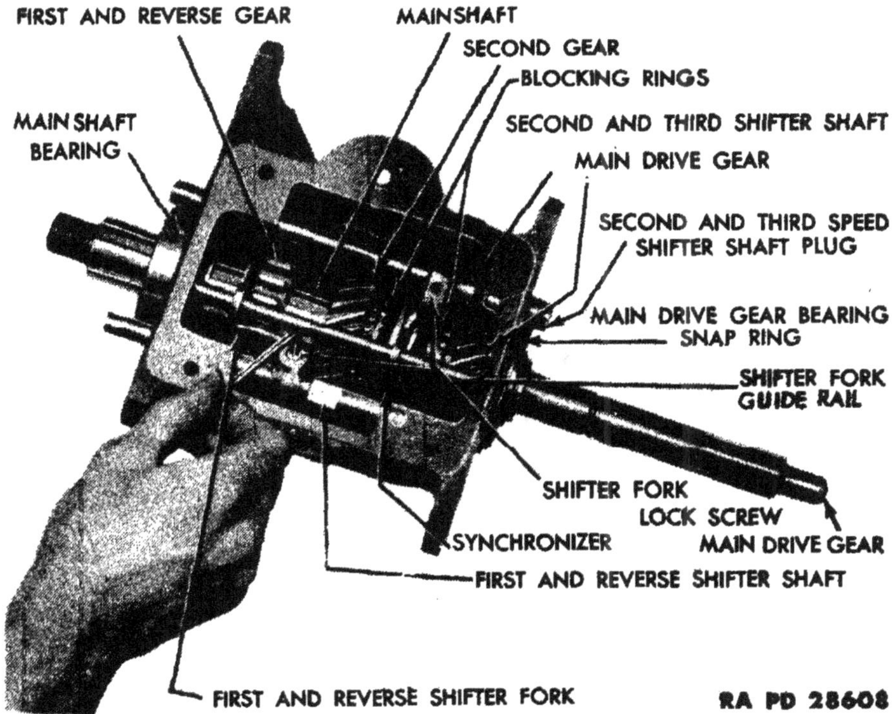

Figure 8 — Removing Shifter Fork Lock Screws

three cap screws from each side of the transmission support crossmember. Place another jack under the transmission. Remove the four bolts that secure the transmission to the clutch housing. Lower both jacks evenly until the transmission support crossmember is approximately 2 inches from the frame. Push the transmission and transfer case to the right so as to free the clutch shaft from the ball joint on the transfer case. Pull the transfer case with transmission straight back until the transmission main drive gear is out of the clutch housing and remove the transfer case and transmission.

j. **Remove Transmission Support Crossmember** (fig. 6). Remove the five mounting bolts that secure the transmission and transfer case to the transmission support crossmember. Remove the transmission support crossmember.

k. **Remove Transmission From Transfer Case** (fig. 27). Drain the oil from the transmission and transfer case. Remove the rear cover from the transfer case. Remove the castellated nut and flat washer that secure the drive gear on the transmission mainshaft and remove the drive gear and oil baffle from the transmission mainshaft, using a suitable puller, if necessary. NOTE: *Vehicles of early manufacture were not supplied with this oil baffle.*

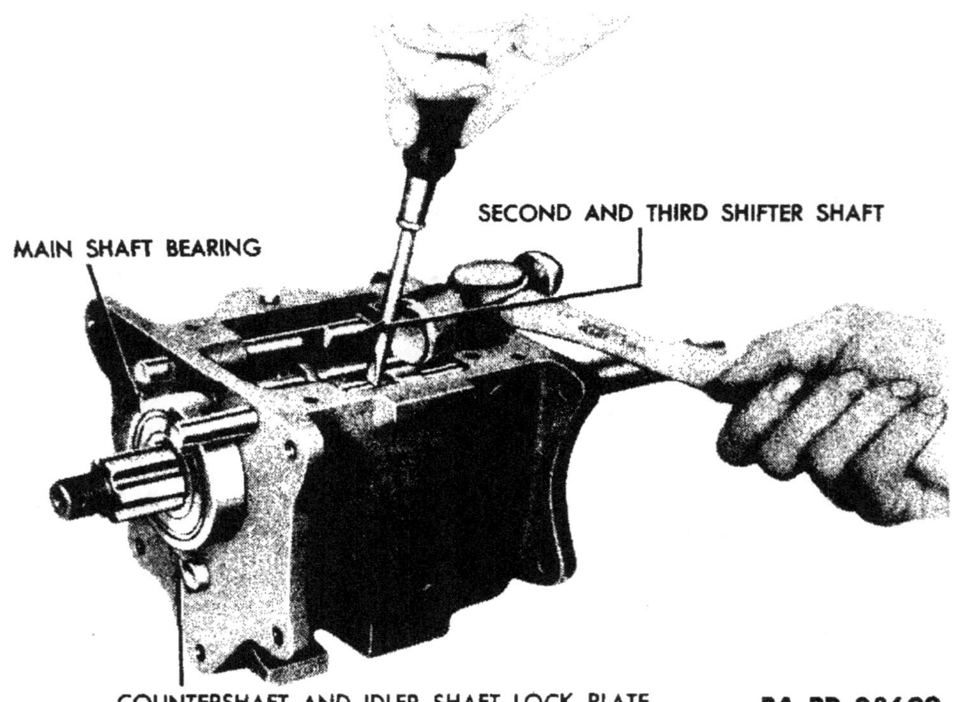

Figure 9 — Removing Shifter Shafts

6. DISASSEMBLY.

a. **Remove Gearshift Housing.** Remove the four cap screws that secure the gearshaft housing to the transmission (fig. 4). Lift the housing, shifter shaft plate, and spring washer from the transmission (fig. 17).

b. **Remove Main Drive Gear Bearing Retainer** (fig. 3). Unhook the clutch release bearing return spring and slide the bearing assembly off the bearing retainer. Remove the three cap screws from the bearing retainer. Slide the bearing retainer and cork gasket off the main drive gear.

c. **Remove Shifter Fork Guide Rail** (fig. 8). Push the shifter fork guide rail out of the transmission.

d. **Remove the Low and Reverse, and the Second and High Shifter Forks.** Remove the shifter fork lock screw from each fork (fig. 8). Tap the shifter shafts part way out of the transmission (fig. 9), being careful not to lose the interlocking ball in each shaft. Hold the shifter fork and pull the shafts from the transmission.

e. **Remove Main Drive Gear.** Tap the countershaft and idle reverse shaft lock plate out of the two shafts (fig. 4). With a long

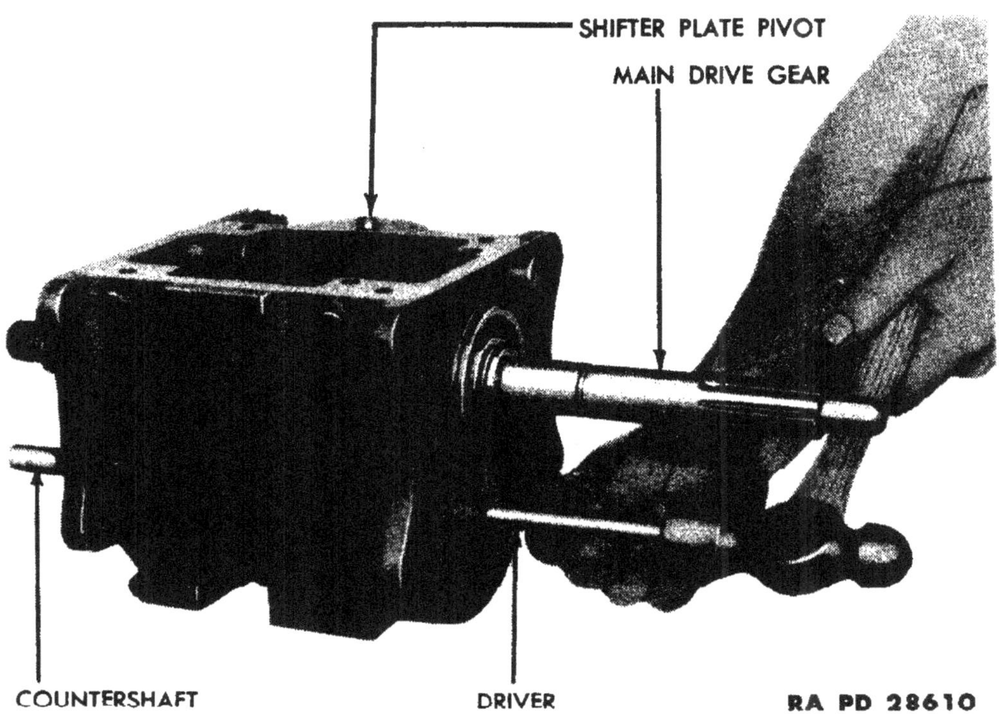

Figure 10 — Removing Countershaft

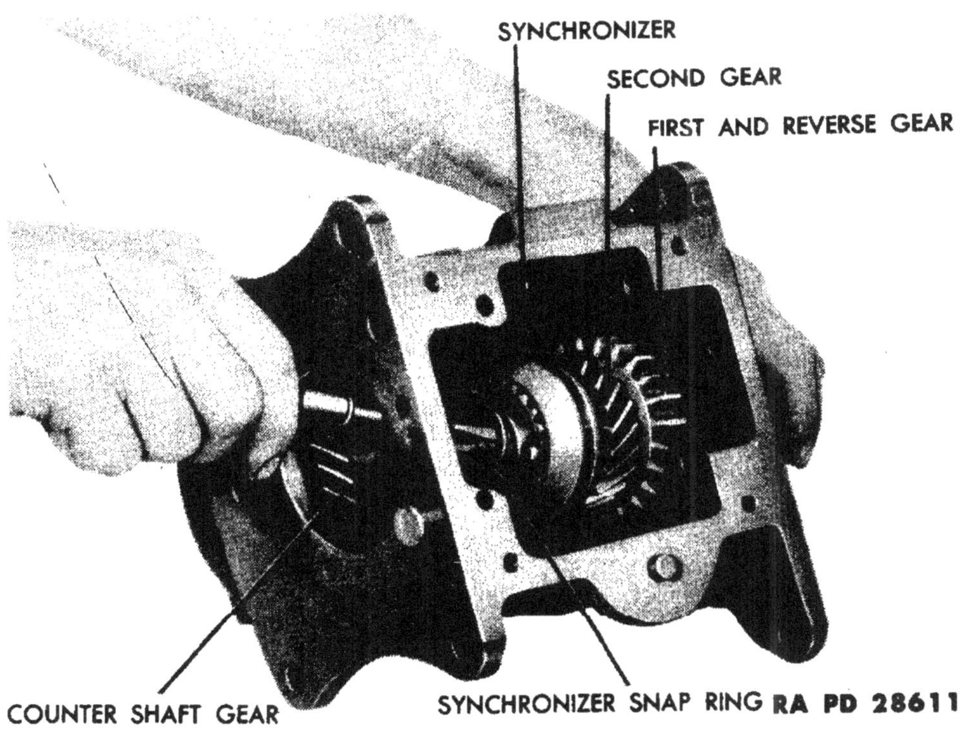

Figure 11 — Removing Synchronizer Hub Snap Ring

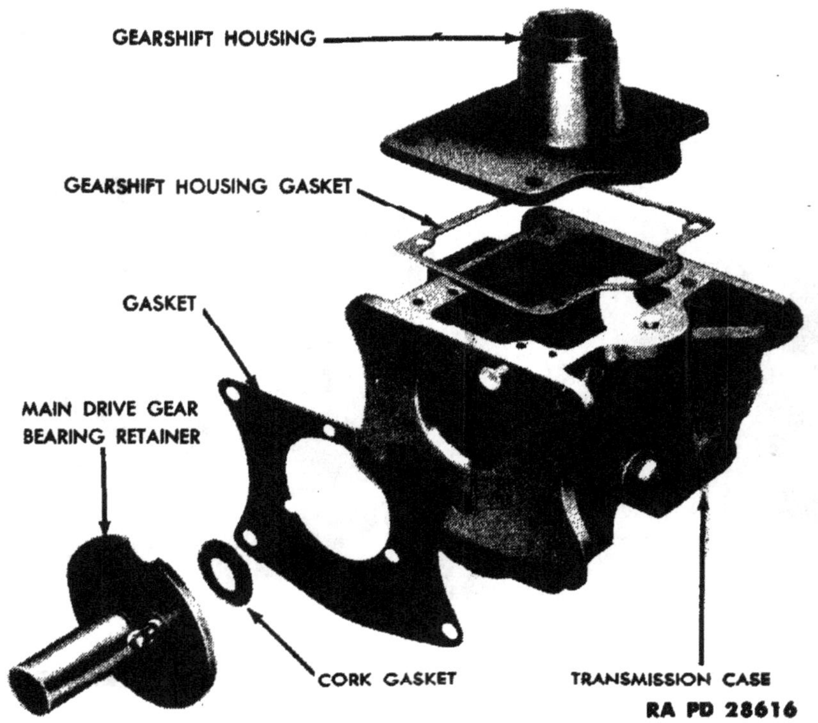

Figure 12 — Transmission Case and Gearshift Housing — Exploded View

drift, tap the countershaft out of the transmission (fig. 10). This will allow the countershaft gear to drop to the bottom of the case for clearance to remove the main drive gear. Pull the main drive gear assembly from the transmission.

f. **Remove Mainshaft** (fig. 11). Remove the synchronizer hub snap ring. Slide the synchronizer assembly, second and first and reverse gear off the mainshaft. Remove the shaft.

g. **Remove Idle Reverse Gear.** Tap the idle reverse gear shaft out of the transmission and remove the gear. Lift the countershaft gear and both thrust washers out of the transmission.

h. **Disassemble Countershaft Gear** (fig. 14). Remove the two bushings and spacer from the countershaft gear.

i. **Disassemble Main Drive Gear** (fig. 13). Remove the snap ring and the 13 rollers from the main drive gear.

j. **Disassemble Synchronizer** (fig. 13). Slide the synchronizer sleeve off the synchronizer hub and remove the two lock rings.

7. CLEANING, INSPECTION, AND REPAIR.

a. **Cleaning.** Wash all parts thoroughly in dry-cleaning solvent until all trace of old lubricant has been removed. Oil the bearings

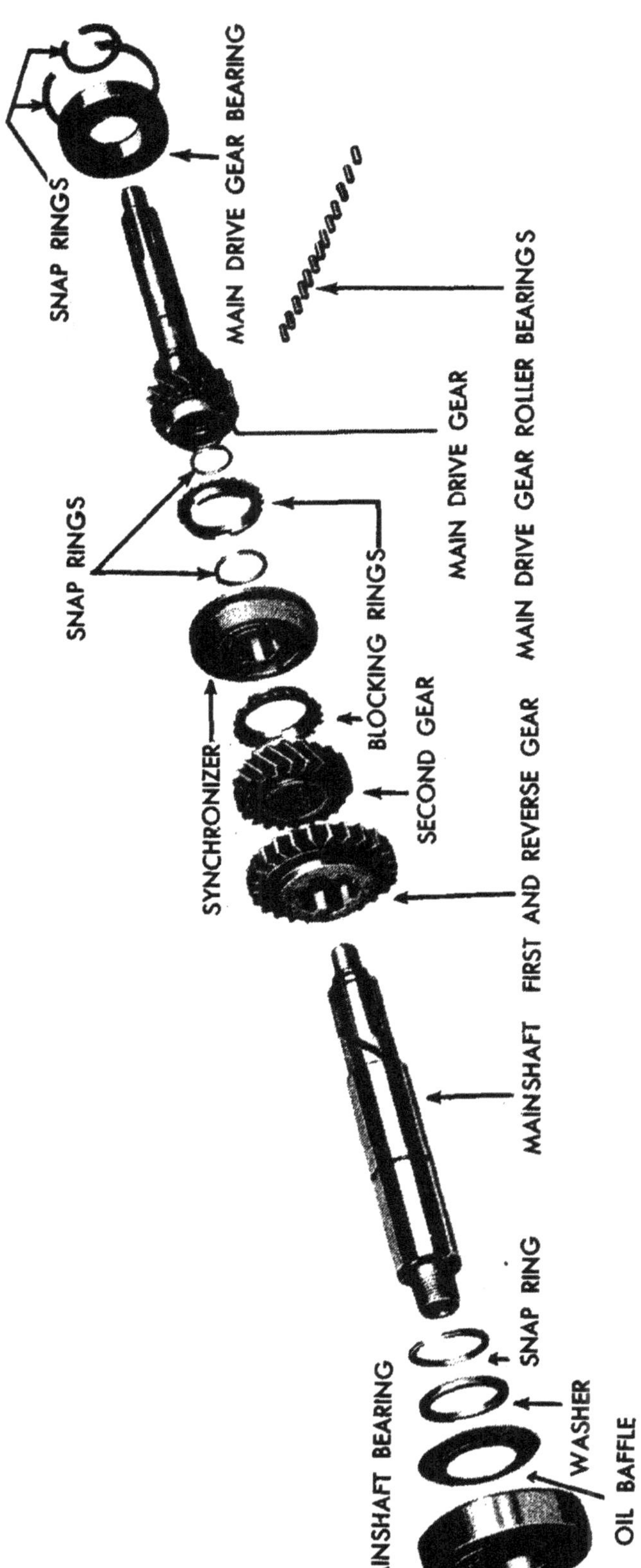

Figure 13 — Mainshaft Assembly — Exploded View

855473 O - 49 - 2

immediately after cleaning to prevent corrosion of the highly polished surfaces.

b. Inspection and Repair.

(1) TRANSMISSION CASE ASSEMBLY (fig. 12). Inspect the case and gearshaft housing for cracks or damage of any kind. Cracked or damaged units must be replaced.

(2) MAIN DRIVE GEAR ASSEMBLY (fig. 13). Replace the main drive gear (clutch shaft) if the following conditions are apparent: Broken teeth or excessive wear; pitted or twisted shaft; discolored bearing surfaces due to overheating. Small nicks can be honed and then polished with a fine stone. Measure the roller bearing recess in the gear end of the shaft. If more than 0.974 inch, replace the main drive gear. Measure the pilot end of the shaft. If it is less than 0.595 inch at the pilot end, replace the main drive gear.

(3) MAINSHAFT (fig. 13). A mainshaft excessively worn, or with pitted or discolored bearing surfaces due to overheating, must be replaced. Measure the diameter of the pilot end of the shaft and the diameter of the second speed gear bearing surface. If they are less than 0.595 inch at the pilot end, or less than 1.126 inches at the second speed gear bearing surface, replace the mainshaft.

(4) FIRST AND REVERSE GEAR (fig. 13). A first and reverse gear with excessively worn teeth or splines, or with broken or chipped teeth must be replaced. Slide the gear onto the mainshaft. If the backlash between the gear and the shaft exceeds 0.005 inch, either the gear or the shaft, or both, must be replaced. A gear with small nicks can be honed and then polished with a fine stone.

(5) SECOND GEAR (fig. 13).

(a) Inspection. A second gear with excessively worn, broken, or chipped teeth, or scored bearing surface must be replaced. Measure the inside diameter of the gear. If more than 1.129 inches the gear bushing must be replaced (step *(b)*, below). Small nicks can be honed and then polished with a fine stone.

(b) Second Gear Bushing Replacement. Place the second gear in an arbor press and, with a suitable driver, press the bushing out of the gear. Use a suitable driver to press a new bushing in the gear. Ream the bushing to from 1.1275 to 1.1280 inches.

(6) COUNTERSHAFT GEAR (fig. 14). Replace excessively worn gears, and gears with broken or chipped teeth, or with pitted or discolored bearing surface due to overheating. Measure the front and rear bearing surfaces of the countershaft gear. If more than 0.7625 inch on either end, replace.

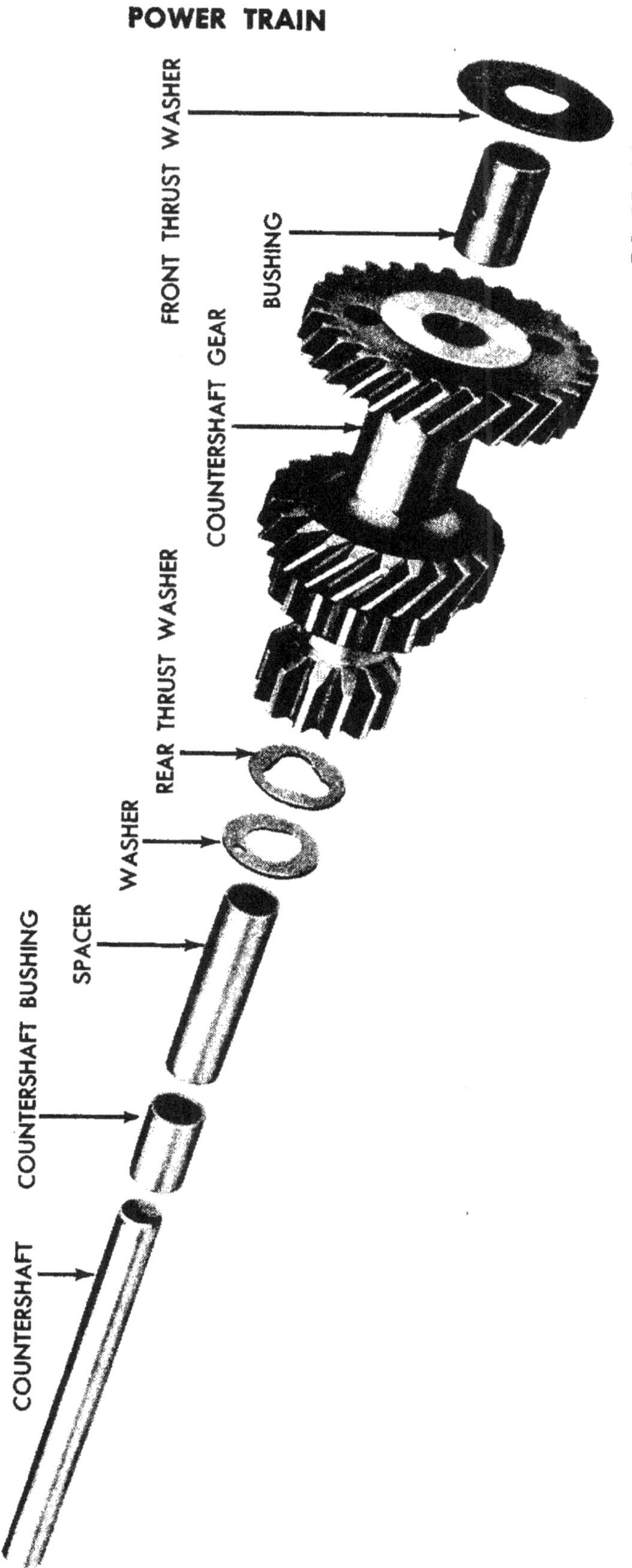

Figure 14 — Countershaft Gear Assembly — Exploded View

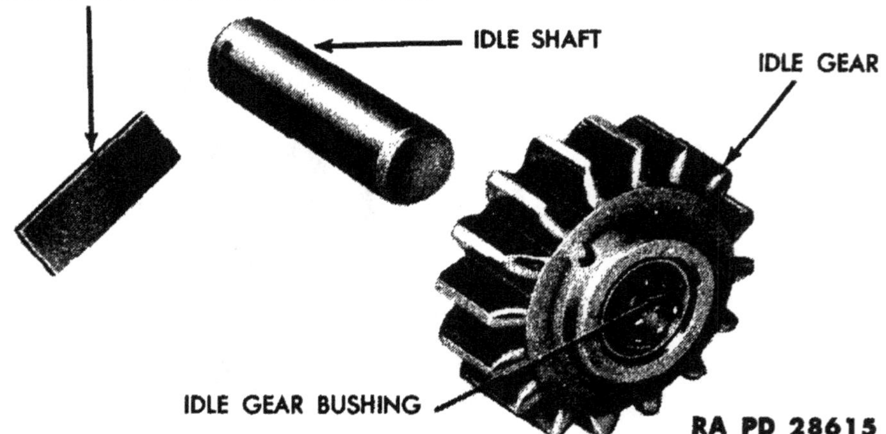

Figure 15 — Idle Gear Assembly — Exploded View

(7) IDLE GEAR (fig. 15).

(a) Inspection. A gear with excessively worn or broken teeth, or with a scored bearing surface must be replaced. Small nicks can be honed and then polished with a fine stone. Measure the inside diameter of the idle gear bushing. If more than 0.626 inch, the bushing must be replaced (step *(b)*, below).

(b) Idle Gear Bushing Replacement. Place the idle gear in an arbor press and, with a suitable driver, press the bushing out of the gear. Use a suitable driver to press a new bushing in the idle gear. Ream the bushing to from 0.623 to 0.624 inch.

(8) IDLE GEAR SHAFT AND COUNTERSHAFT (figs. 14 and 15). Ridged, scored, or excessively worn, shafts must be replaced. An idle gear shaft measuring under 0.6185 inch or countershaft measuring under 0.7490 inch must be replaced.

(9) SYNCHRONIZER (fig. 13). Blocking rings with worn, broken, or nicked teeth, must be discarded. Hubs with excessively worn splines must be replaced. Sleeves with broken, nicked, or worn teeth, or excessively worn splines, must be replaced.

(10) MAIN DRIVE GEAR BEARING ROLLERS (fig. 13). Needle bearing rollers with flat spots, pitted, or discolored surfaces must be replaced. Measure the diameter of each roller. If less than 0.187 inch, the rollers must be replaced.

(11) BALL BEARINGS (fig. 13). Ball bearings with loose or discolored balls, or with pitted or cracked races must be replaced.

(12) COUNTERSHAFT THRUST WASHERS (fig. 14). Replace excessively worn or ridged thrust washers. Measure each thrust wash-

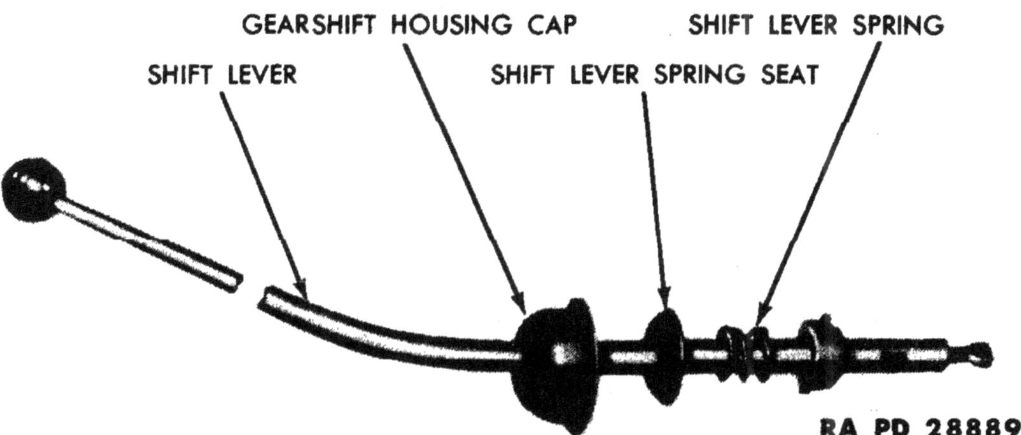

Figure 16 — Transmission Shift Lever

er. If the front washer is less than 0.029 inch, or if either of the rear washers are less than 0.060 inch, they must be replaced.

(13) COUNTERSHAFT BUSHINGS (fig. 14). Excessively worn, scored, or ridged countershaft bushings must be replaced. Measure the inside and outside diameter of the bushings. If the outside diameter is less than 0.759 inch, or if the inside diameter is more than 0.6225 inch, the bushings must be replaced.

(14) SHIFT LEVER (fig. 16). Replace the shift lever if it is excessively worn or bent. Check the gearshift housing cap for stripped threads. Replace the shift lever spring, if it is cracked.

8. ASSEMBLY.

a. Install Idle Gear. Hold the idle gear (fig. 15) in place in the case with the cone end of the hub toward the front, and push the idle gear shaft into the case.

b. Install Countershaft Gear (fig. 14). Dip the countershaft bearings into SAE 90 oil. Slide the spacer into the countershaft gear and install a bushing in each end of the countershaft gear. Coat the front thrust washer, rear thrust washer, and steel washer with a light film of grease to hold them in place while installing the gear. Lay the countershaft gear in the case with the large gear toward the front.

c. Install Mainshaft Assembly (fig. 13). Insert the mainshaft in the case through the opening in the rear of the case. Slide the first and reverse gear onto the shaft, with the shifter fork channel toward the rear. Slide the second gear onto the mainshaft with the tapered end of the gear toward the front. Install a blocking ring onto the second gear. Slide the synchronizer onto the mainshaft with the long end of the hub toward the front and install the snap ring.

d. Install Main Drive Gear Assembly (fig. 13). Place the other blocking ring in the synchronizer and install the main drive gear assembly in the case.

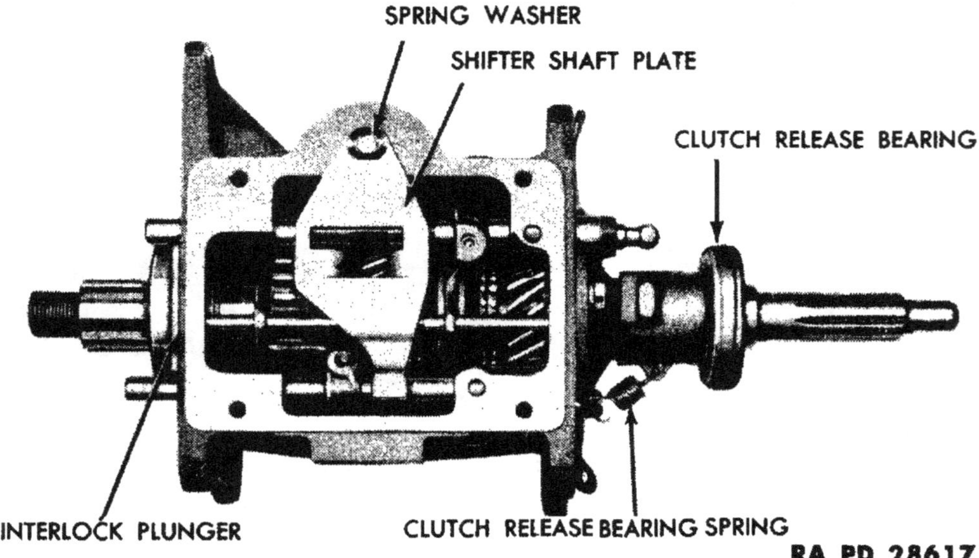

Figure 17 — Gears Installed in Transmission — Top View

e. **Install Countershaft.** Raise the countershaft gear into position. Making sure the three washers are in line, push the countershaft into the case and tap the lock plate between the countershaft and idle gear shaft (fig. 4).

f. **Install First and Reverse Shifter Fork** (fig. 8). Hold the first and reverse shifter fork in position on the first and reverse gear, and slide the low and reverse shifter shaft (short shaft) into the case about half way. Drop an interlock spring and ball in the pocket. Press down on the ball and push the shifter shaft all the way in the case. Line up the groove of the shaft with the shifter fork and install the lock screw.

g. **Install Second and Third Shifter Fork** (fig. 8). Repeat the same procedure as used in installing the low and reverse shifter fork, and then push the guide rail into the case and through both shifter forks.

h. **Install Gearshift Housing on Case** (fig. 17). Place the transmission in neutral position. Lay the shifter shaft plate on the pivot and on the shifter shafts. Lay the spring washer on the pivot. Place a new gearshift housing gasket on the case. Place the shift lever in neutral position. Lay the housing on the transmission and install the four lock washers and cap screws in the housing.

i. **Install Clutch Release Bearing** (fig. 3). Slide the clutch release bearing assembly onto the main drive gear bearing retainer and install the clutch release bearing return spring.

9. INSTALLATION.

a. **Install Transmission to Transfer Case.** Place the transmission in position on the transfer case. Be sure the interlock plunger (fig. 4) is in position between the two shifter shafts on the transmission. Install the bolts that secure the transmission to the transfer case. Slide the oil baffle and mainshaft gear on the transmission mainshaft through the rear cover opening on the transfer case. (The oil baffle was not supplied on vehicles of early manufacture. If grease is found to have been leaking from the transfer case into the transmission on vehicles without this baffle, reverse the rear mainshaft bearing (fig. 13) so that the open side of the bearing faces the front of the transmission. Leave the oil baffle in front of the bearing in its original position. Install another oil baffle at the rear of the bearing.) Install the flat washer and nut that secure the mainshaft gear to the transmission mainshaft. Install a new gasket and the rear cover on the transfer case (fig. 27).

b. **Place Transmission in Position on Vehicle.** Place a jack under the transmission and raise the transmission and transfer case up until the shaft of the main drive gear is lined up with the splines in the clutch disk.

c. **Install Transmission Main Drive Gear to Clutch Housing.** Insert the shaft of the main drive gear into the clutch splines carefully, do not use force. Slide the transmission in flush with the clutch housing. Install the four bolts that secure the transmission to the clutch housing.

d. **Install Clutch Shaft to Transfer Case** (fig. 6). Push the transfer case to the right until the clutch shaft has enough clearance to enter the ball joint on the transfer case.

e. **Install Transmission Support Crossmember** (fig. 6). Place the transmission support crossmember in position on the transmission. Install the four bolts that secure the crossmember to the transmission. Raise the transmission up with a jack until the crossmember is flush with the frame. With a long nosed drift, line up the holes on the crossmember with the holes in the frame. Install the three nuts and bolts on each end of the crossmember and remove the jack. Install the transfer case mounting bolt.

f. **Install Clutch Release Fork** (fig. 7). Working through the inspection plate opening on the clutch housing, insert the clutch release fork in the clutch housing. Place the release fork behind the clutch release bearing. Slide the clutch release fork cable in the slot on the opposite end of the clutch release fork. Install the clutch release fork cable to the clutch shaft at the transfer case.

g. **Install Hand Brake Cable** (fig. 6). Install the hand brake cable to the brake band at the transfer case. Install the hand brake spring leading from the brake band linkage to the body floor plate. Install the clamp that secures the hand brake cable to the transfer case.

h. **Install Engine Stay Cable and Ground Strap** (fig. 6). Install the engine stay cable leading from the engine rear plate to the transmission support crossmember. Install the ground strap leading from the transmission to the floor plate.

i. **Install Propeller Shafts and Speedometer Cable** (fig. 6). Install the rear propeller shaft to the transfer case (par. 21 a). Install the front propeller shaft to the transfer case (par. 21 b). Install the speedometer cable to the transfer case.

j. **Install Transmission Shield** (fig. 6). Install the five nuts and bolts that secure the shield to the transmission support crossmember. Install the clamp that secures the exhaust pipe to the shield.

k. **Lubricate and Adjust Clutch.** Fill both the transmission and transfer case to proper oil level with specified oil. Adjust the clutch pedal free travel (refer to TM 9-803).

Section III

TRANSFER CASE

10. DESCRIPTION AND DATA.

a. **Description.** The transfer case (figs. 28 and 29) is located at the rear of the transmission. The transfer case is essentially a 2-speed transmission, which provides two gear ratios and a means of distributing the power from the transmission to the two axles.

b. **Data.**

Make .... Spicer
Model 18
Mounting .... Unit with transmission
Shift lever .... Floor
Ratio:
  High .... 1 to 1
  Low .... 1.97 to 1

Bearings:

Transmission mainshaft	Ball
Idle gear	2 rollers
Output shaft	Taper rollers
Front axle clutch shaft front bearing	Ball
Rear pilot in output shaft	Bronze bushing

11. REMOVAL.

a. **Remove Transmission Shield** (fig. 6). Remove the two cap screws that secure the exhaust pipe clamp to the shield. Remove the exhaust pipe clamp. Remove the five bolts that secure the transmission shield to the transmission support crossmember and remove the shield.

b. **Remove Hand Brake Cable and Clutch Cable** (fig. 6). Remove the hand brake spring at the transfer case. Remove the clevis pin that secures the hand brake cable at the brake on the transfer case. Remove the hand brake cable clamp on the transmission. Remove the clevis pin from the clutch cable at the transmission support crossmember.

c. **Remove Mounting Bolt and Rear Cover** (figs. 7 and 27). Remove the mounting bolt that secures the transfer case to the transmission support crossmember at the right side of the transfer case. Remove the five cap screws that secure the rear cover to the transfer case.

d. **Remove Rear Propeller Shaft** (fig. 7). Disconnect the rear propeller shaft at the transfer case (par. 17 b).

e. **Remove Mainshaft Gear** (fig. 27). Through the opening at the rear of the transfer case, remove the castellated nut that secures the mainshaft gear to the transmission mainshaft. Remove the flat washer mainshaft gear and oil retainer.

f. **Remove Transfer Case.** Place a jack under the transfer case. Remove the five cap screws that secure the transfer case to the transmission. Move the transfer case straight back until the transmission mainshaft is out of the transfer case. Remove the transfer case.

12. DISASSEMBLY.

a. **Remove Brake Band and Drum Assembly** (fig. 28). Remove the two anchor screws from the brake band. Remove the brake band adjusting nut and adjusting screw. Remove the clevis pin from the hand brake linkage. Remove the brake band assembly. Remove the castellated nut that secures the universal joint flange to the output shaft. Install puller 41-P-2912 on the universal joint flange and remove the flange and brake drum (fig. 18). NOTE: *The puller illustrated in figure 18 is similar to puller 41-P-2912.*

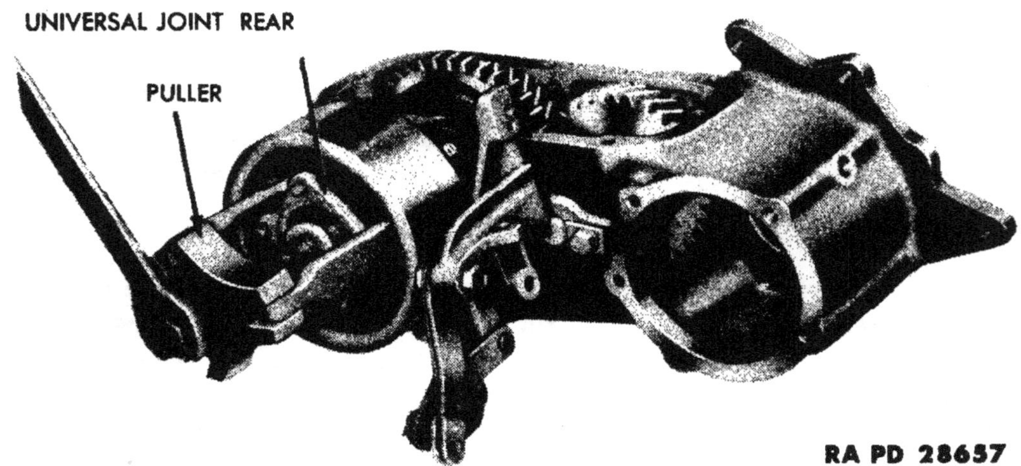

Figure 18 — Removing Rear Universal Joint Flange With Puller Similar to Puller 41-P-2912

b. Remove Rear Output Shaft Bearing Cap (fig. 26). Remove the four cap screws that secure the rear output shaft bearing cap to the transfer case housing. Remove the rear output shaft bearing cap. Remove the rear bearing cap shims. Remove the speedometer drive gear from the output shaft.

c. Remove Intermediate Gear and Bottom Cover (figs. 25 and 27). Remove the 10 cap screws that secure the bottom cover to the transfer case and remove the bottom cover. Remove the cap screw that secures the lock plate. Remove the lock plate. With a suitable driver, remove the intermediate gear shaft. Remove the intermediate gear, thrust washers, and roller bearings through the bottom of the transfer case.

d. Remove Shifter Shaft and Front Output Shaft Bearing (fig. 29). Shift front axle drive to the engaged position. Remove the poppet plug, spring, and ball on both sides of the output shaft bearing cap. Remove the five cap screws that secure the front output shaft bearing cap to the transfer case. Remove the front output shaft bearing cap as an assembly with the universal joint flange, clutch shaft, bearing, clutch gear, shifter fork, and shifter rod. Be careful not to lose the interlock in the front bearing cap.

e. Remove Output Shaft (fig. 19). Insert a screwdriver between the snap ring and output shaft bearing and pry the output shaft bearing away from the snap ring. Remove the snap ring from the groove in the output shaft. Pull the output shaft out from the rear of the housing. The output shaft bearing, snap ring thrust washer, output shaft sliding gear, and output shaft gear can now be removed through the bottom of the transfer case.

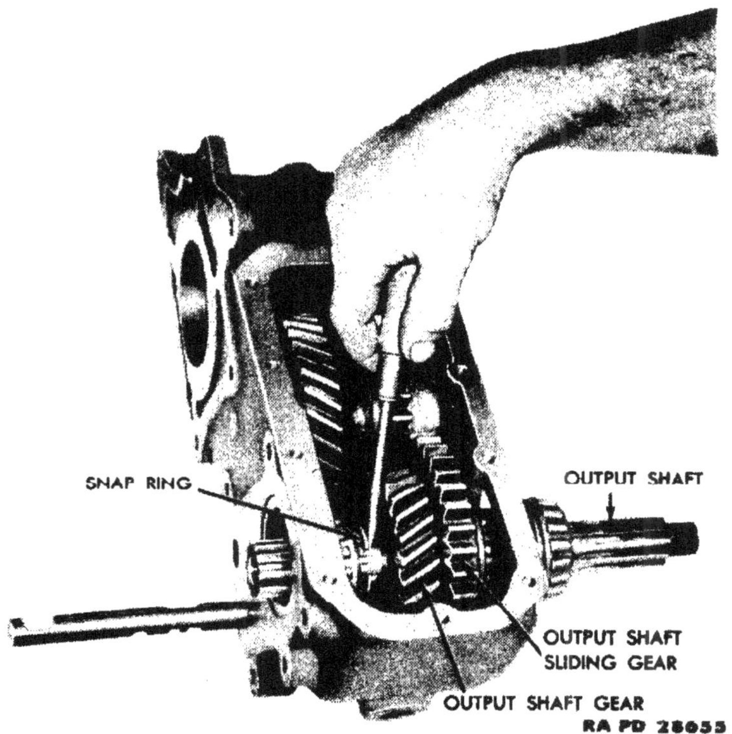

Figure 19 — Removing Snap Ring From Output Shaft

f. Disassemble Front Output Shaft Bearing Cap (fig. 21). Remove the set screw that secures the shifter fork to the front wheel drive shifter shaft. Slide the shifter shaft out of the shifter fork. Remove the shifter fork and clutch gear from the bearing cap. Remove the snap ring that secures the output shaft bearing and remove the output shaft bearing from the bearing cap.

13. CLEANING, INSPECTION, AND REPAIR.

a. Cleaning. Cleaning all parts thoroughly in dry-cleaning solvent. Clean the bearings by rotating them while immersed in dry-cleaning solvent until all trace of lubricant has been removed. Oil the bearings immediately to prevent corrosion of the highly polished surface.

b. Inspection.

(1) TRANSFER CASE ASSEMBLY (fig. 27). Inspect the transfer case housing for cracks or damage of any kind. Inspect the bottom and rear cover for bent or damaged condition. Replace the gaskets on the bottom and rear covers.

(2) FRONT OUTPUT SHAFT BEARING CAP ASSEMBLY (fig. 21).

(a) Front Output Shaft Bearing Cap Housing (fig. 20). Replace the front bearing cap, if it is cracked or damaged. Shifter shaft and output shaft oil seals must be replaced (subpar. c, below).

Figure 20 — Front Output Shaft Bearing Cap Housing and Oil Seals

(b) Front Wheel Drive Shifter Shaft and Fork (fig. 21). Replace the front wheel drive shifter shaft, if bent or damaged. Replace the fork if it has stripped set screw threads, if it is cracked or has bent forks.

(c) Clutch Shaft and Gear (fig. 21). Replace the clutch shaft if the splines or gear teeth are chipped or worn, if the gear has any teeth missing. Check the diameter of the pilot end of the clutch shaft. If the diameter is less than 0.625 inch, replace the clutch shaft. Replace the clutch gear, if it is worn or has any broken teeth.

(d) Output Shaft Bearing (fig. 21). Ball bearings with loose or discolored balls or with pitted or cracked races must be replaced.

(3) INTERMEDIATE GEAR ASSEMBLY (fig. 25). Replace the intermediate gear if excessively worn, or if any teeth are damaged. Check the thickness of the thrust washers. If the thrust washers are less than 0.093 inch in thickness, replace them. Check the diameter of the intermediate gear shaft. If the diameter is less than 0.750 inch, replace the intermediate gear shaft. Replace the roller bearing, if the rollers are scored or have flat spots.

(4) REAR OUTPUT SHAFT BEARING CAP ASSEMBLY (fig. 26). Replace the output shaft bearing cap if cracked or damaged. Replace the speedometer drive gear if it is worn or has damaged teeth. Replace the oil seal in the output shaft bearing cap housing (subpar. c, below). Replace the brake drum if it is worn or bent. Replace the universal joint rear flange, if the splines are worn. Replace the dust shield on the flange if bent.

(5) OUTPUT SHAFT ASSEMBLY (fig. 24). Replace the output shaft if the splines are worn. Small nicks can be removed by honing and then polishing with a fine stone. Measure the inside diameter of

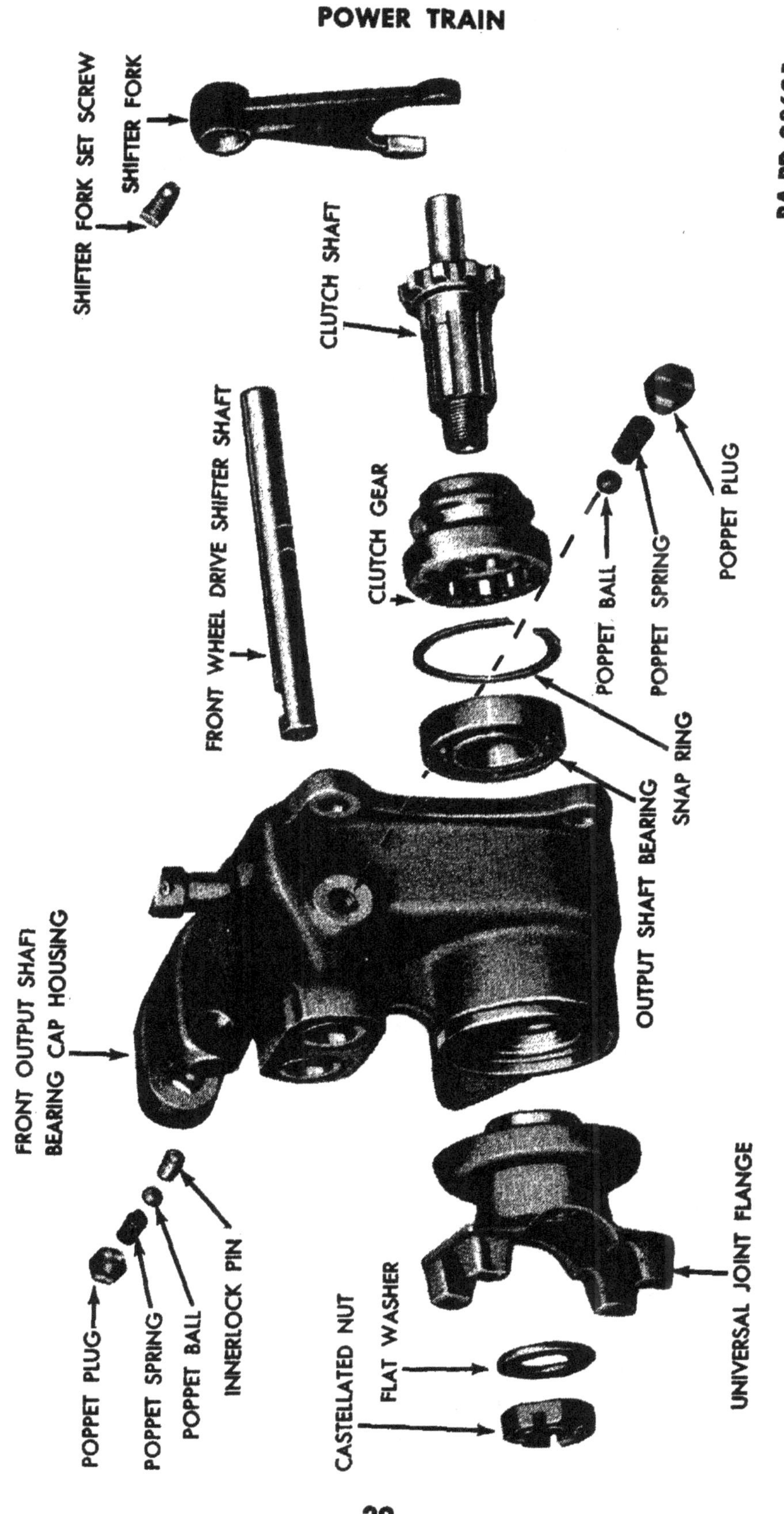

Figure 21 — Front Output Shaft Bearing Cap — Exploded View

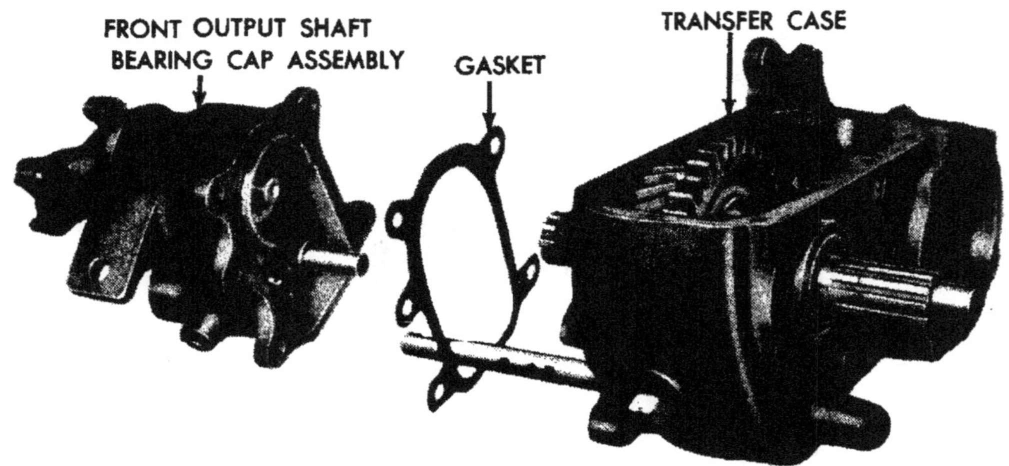

Figure 22 — Installing Front Output Shaft Bearing Cap to Transfer Case

the bushing in the output shaft. If it is greater than 0.627 inch, replace the output shaft. Replace the output shaft gear if it is worn or has any damaged teeth. Replace the sliding gear, if it is worn or has damaged teeth. Measure the thickness of the thrust washer. If the thrust washer thickness is less than 0.103 inch, replace it. Replace the roller bearings if they are scored or have flat spots, or if the races are nicked or cracked.

(6) UNDER DRIVE SHIFTER FORK ASSEMBLY (fig. 24). Check the fork for stripped set screw threads, cracked or bent forks. Replace if in any of these conditions. Replace the under drive shifter shaft if it is bent.

(7) SHIFT LEVER ASSEMBLY (fig. 29). Replace the shift levers if found bent or damaged. Replace the shift lever spring if bent or cracked. Measure the diameter of the shift lever pivot pin. If the diameter is less than 0.500 inch, replace the pivot pin.

c. **Output Shaft Bearing Cap Oil Seal Replacement** (fig. 20). Drive the old oil seal out of the output shaft bearing cap housing, using a suitable driver. Drive the oil seals out, working from the inside of the cap housing. To install a new oil seal, use a driver the size of the oil seal and drive the new seal in the output shaft bearing cap housing.

14. ASSEMBLY.

a. **Assemble the Front Output Shaft Bearing Cap** (fig. 21). Insert the bearing in the output shaft bearing cap. Install the snap ring that secures the bearing in the output shaft bearing cap. Insert the clutch shaft through the bearing from the inside of the output shaft bearing cap. Insert the front wheel drive shifter shaft in the output

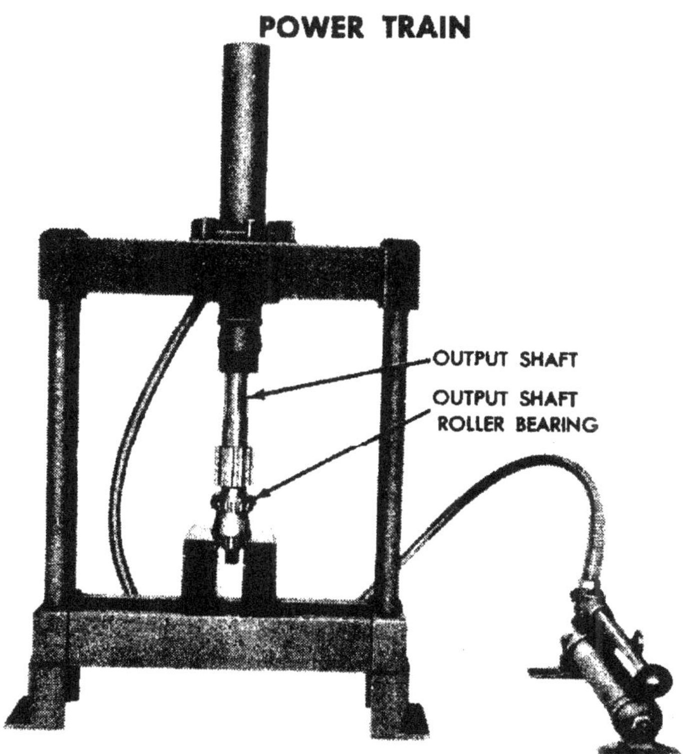

Figure 23 — Pressing Output Shaft Bearing on Output Shaft

shaft bearing cap through the outer side of the output shaft bearing cap. Place the front wheel drive shifter fork in position on the clutch gear. Slide the shifter fork on the shifter shaft and clutch gear on the clutch shaft together. Install the set screw in the shift fork and secure with a lock wire. Install the universal joint flange on the clutch shaft. Install the washer and castellated nut that secure the universal joint flange to the clutch shaft.

b. Install Under Drive Shifter Fork (fig. 20). Place the under drive shifter fork in the transfer case housing. Insert the under drive shifter shaft in the transfer case and shifter fork. Install the shifter fork set screw that secures the fork to the shifter shaft. Secure the set screw with lock wire.

c. Install Output Shaft in Transfer Case (figs. 23 and 24). Press the rear output shaft bearing on the output shaft (fig. 23). Set the output shaft sliding gear in the transfer case with the shifter fork in the channel of the sliding gear. Place the output shaft gear in the transfer case with the shoulder of the output shaft gear facing the sliding gear. Insert the output shaft in the transfer case and through the gears. Slide the thrust washer on the output shaft. Install the snap ring that secures the output shaft gear on the shaft. Slide the front output shaft roller bearing on the output shaft and, using a suitable driver, tap the roller bearing snug against the snap ring. Tap the front roller bearing cup

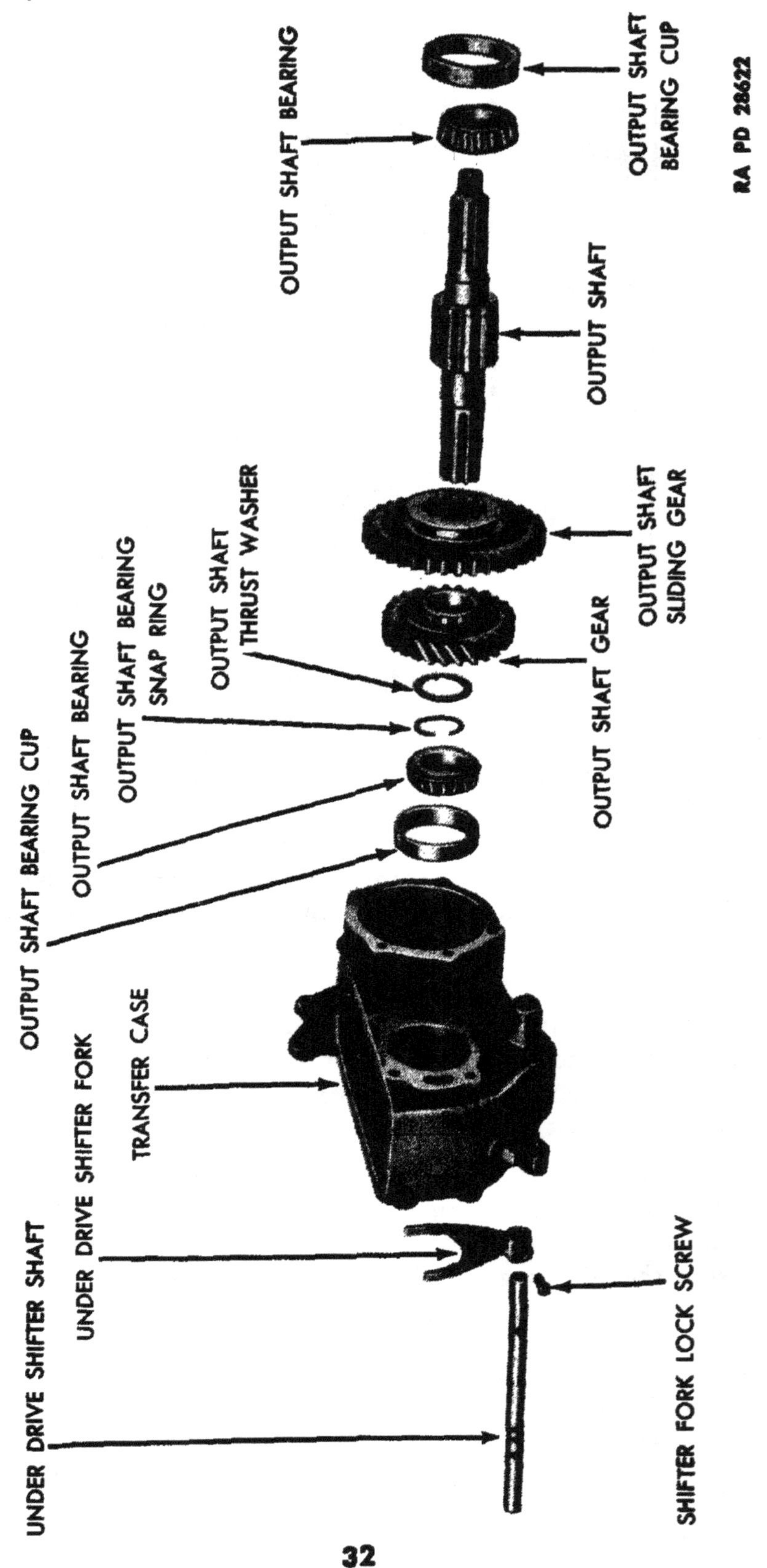

Figure 24 —Output Shaft — Exploded View

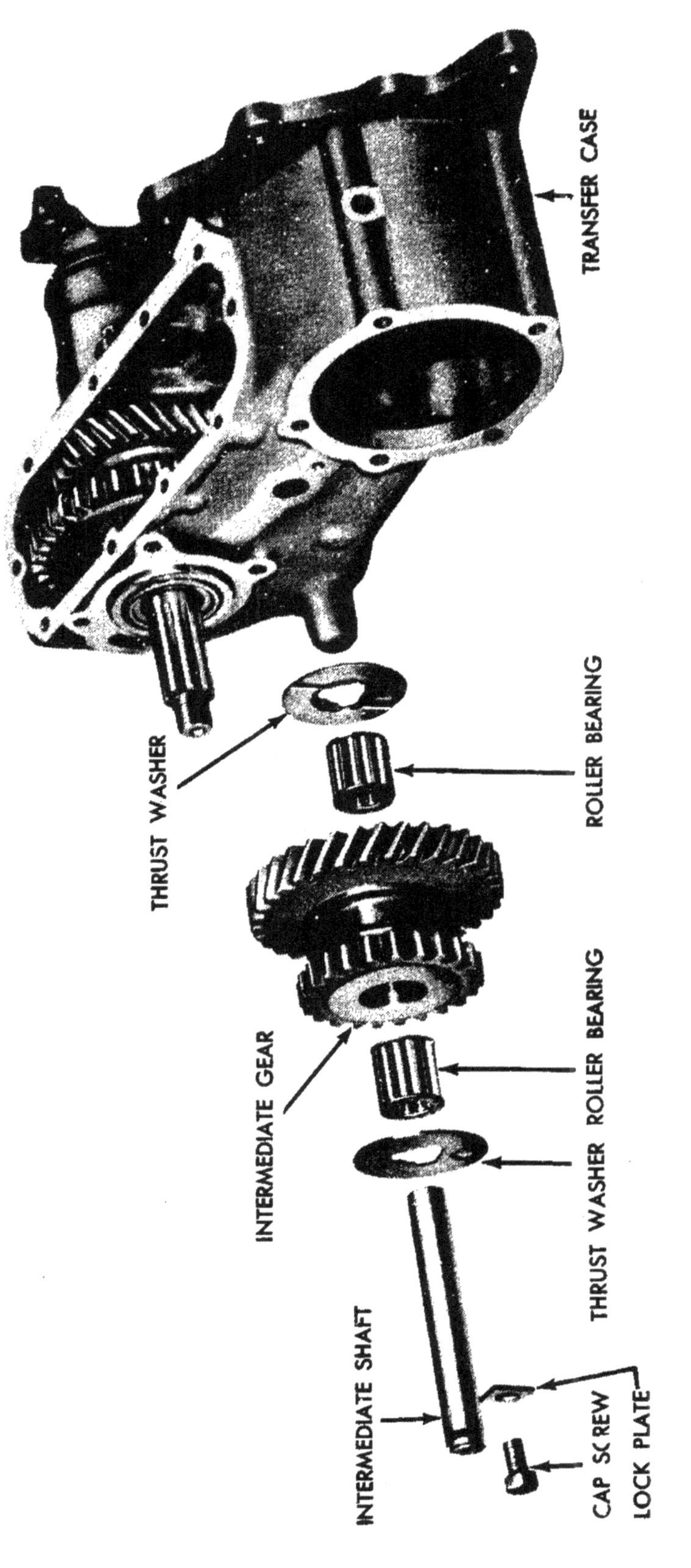

Figure 25 — Intermediate Gear Assembly — Exploded View

855473 O - 49 - 3

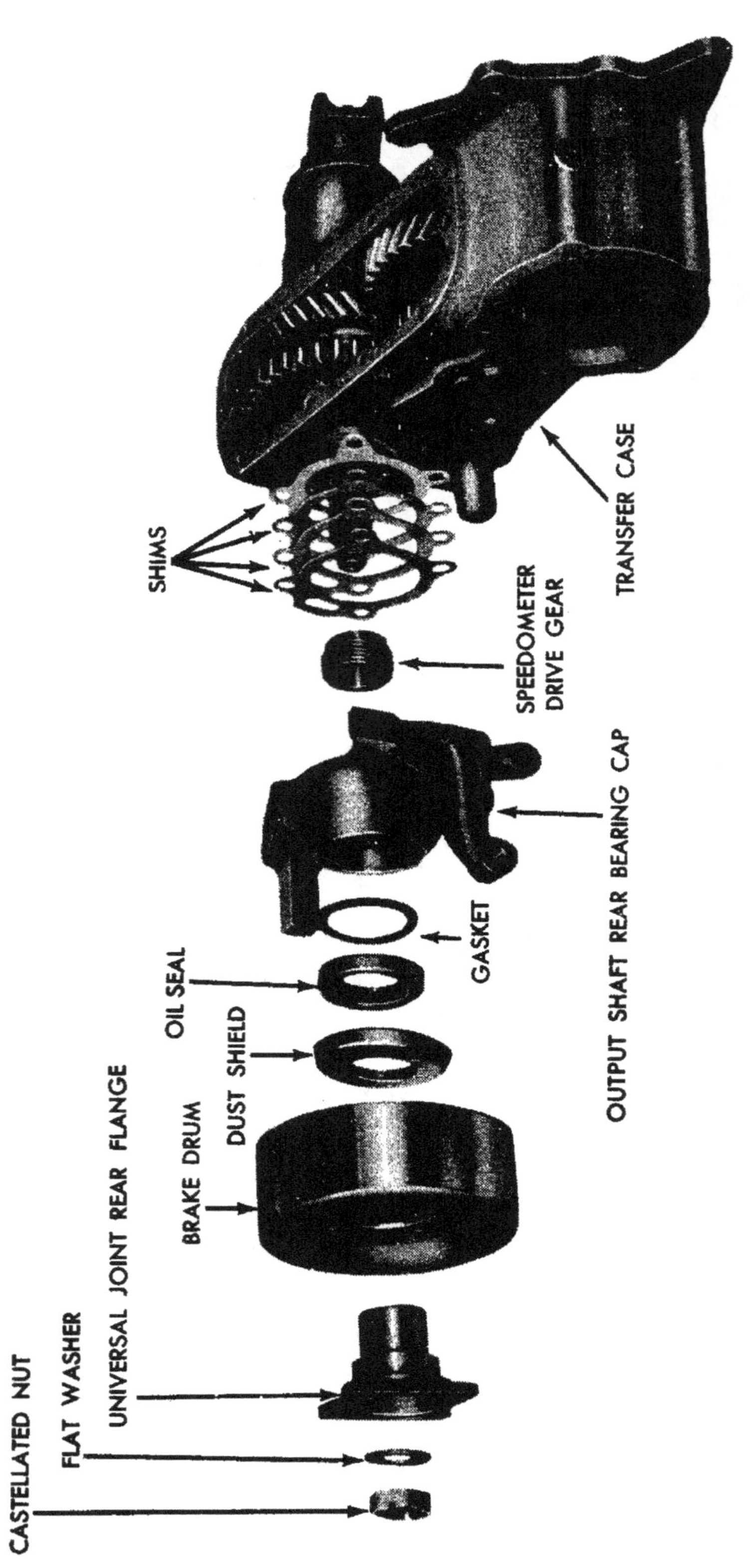

Figure 26 — Rear Output Shaft Cap — Exploded View

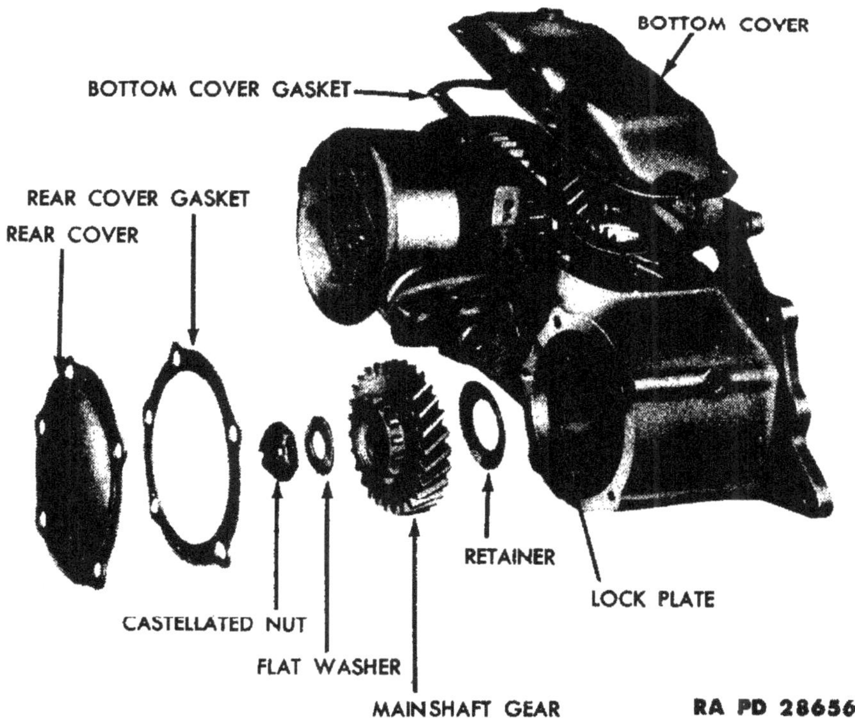

Figure 27 — Bottom Cover and Mainshaft Gear — Exploded View

in the transfer case until the cup is slightly below flush with the transfer case. Tap the rear bearing cup in the transfer case until the cup is approximately 1/8 inch from the transfer case surface.

d. **Install Front Output Shaft Bearing Cap to Transfer Case** (figs. 21 and 22). Place a new gasket in position on the transfer case. Install the interlock (fig. 21) in the interlock opening on the bearing cap. Slide the front output shaft bearing cap on the under drive shifter shaft, being careful not to damage the oil seal in the output shaft bearing cap. Install the five bolts that secure the front bearing cap to the transfer case. Install the poppet ball, poppet spring and poppet plug on both sides of the front bearing cap (fig. 21).

e. **Install Intermediate Gear** (fig. 25). Insert the roller bearings in the intermediate gear. Place the thrust washers in the transfer case, with the side having the bronze facing, toward the intermediate gear. Apply grease to the thrust washers to hold them in position, if necessary. Place the intermediate gear between the thrust washers in the transfer case. Install the intermediate gear shaft in the transfer case. Install the lock plate that secures the intermediate gear shaft to the transfer case.

f. **Install Rear Output Shaft Cap to Transfer Case** (fig. 26). Slide the speedometer drive gear on the output shaft. Install the oil seal in

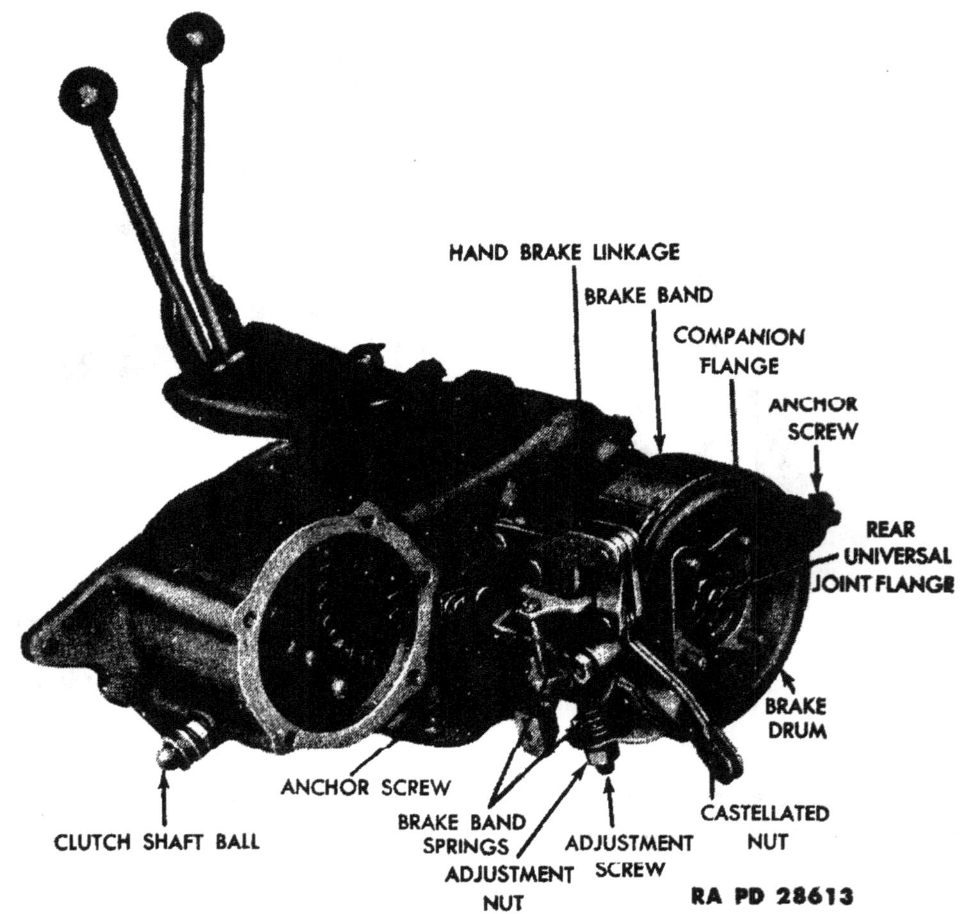

Figure 28 — Transfer Case

the rear output shaft cap (par. 13 c). Install the rear output shaft cap, shims and gasket on the transfer case. Tighten the four cap screws evenly to prevent cracking the output shaft cap. Shims are to be added or removed until the output shaft has no end play, but turns freely. When adjusting the bearings, each time shims are added, the shaft must be free before attempting to tighten the output shaft cap again. Insert the rear universal joint flange in the brake drum. Place the four cap screws in the brake drum and universal joint flange, using a suitable driver, drive the dust shield on the universal joint flange. Install the rear universal joint flange on the output shaft, and install the flat washer and nut.

g. **Install Bottom Cover to Transfer Case** (fig. 27). Install a new gasket in position on the transfer case. Place the bottom cover on the transfer case. Install the cap screws that secure the bottom cover to the transfer case.

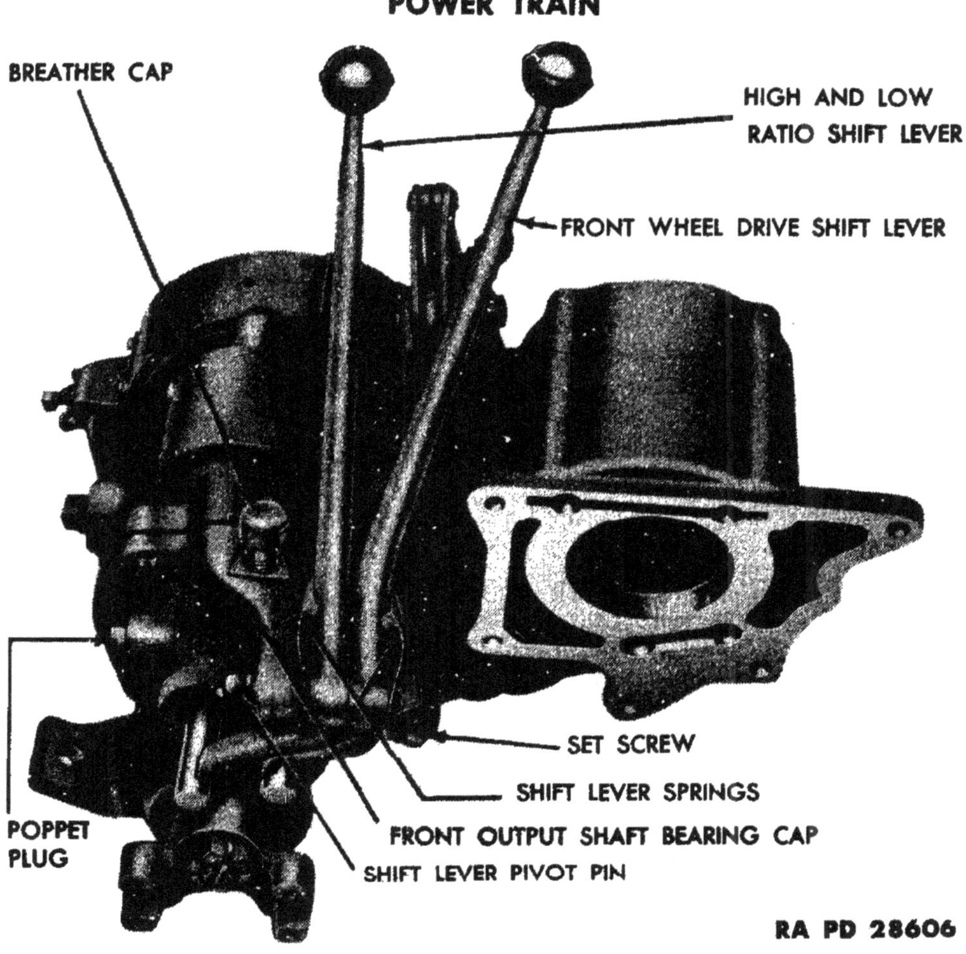

Figure 29 — Transfer Case Shift Levers

h. Install Brake Band to Transfer Case (fig. 28). Place the brake band on the brake drum. Place the brake band springs between the rear output shaft bearing cap and the ends of the brake band. Install the nut and bolt that secure the hand brake linkage to the rear output shaft bearing cap. Insert the adjusting screw through the brake band linkage, brake band springs, and install the adjusting nut. Install the two anchor screws on the brake band.

15. INSTALLATION.

a. Raise Transfer Case. Raise the transfer case and line up the clutch shaft ball joint in the transfer case. Line up the transfer case with the transmission. Be sure the interlock is in position on the rear of the transmission case before installing the transfer case to the transmission (fig. 4). Install the five cap screws that secure the transfer case to the transmission. Install the mounting bolt that secures the transfer case to the transmission support crossmember.

b. Install Mainshaft Gear (fig. 27). Insert the retainer and mainshaft gear on the transmission mainshaft. Install the flat washer and castellated nut that secure the mainshaft gear on the transmission mainshaft. Place a new gasket and the rear cover on the transfer case and install the cap screws that secure the cover to the case.

c. Install Clutch, Hand Brake and Speedometer Cables (fig. 6). Install the clevis that secures the clutch release fork cable to the clutch shaft. Install the clevis pin that secures the hand brake cable to the brake band. Install the cap screw that secures the hand brake clamp to the transfer case rear output shaft cap. Install the speedometer cable to the transfer case at the top of the rear output shaft cap.

d. Install Propeller Shaft and Transfer Case Shield (fig. 6). Connect the rear propeller shaft to the transfer case (par. 17 **b**). Place the transmission shield in position and install the five cap screws that secure the shield to the transmission support crossmember. Install the exhaust pipe clamp to the transmission shield. Fill the transfer case with specified oil to the proper level. Adjust the hand brake band (refer to TM 9-803).

Section IV

PROPELLER (DRIVE) SHAFTS AND UNIVERSAL JOINTS

16. DESCRIPTION AND TABULATED DATA.

a. Description (fig. 2). The power from the transfer case is carried through two propeller shafts. One propeller shaft runs from the front of the transfer case to the front axle, and a second propeller shaft runs from the rear of the transfer case to the rear axle. Each is equipped with two universal joints. The splined slip joint at one end of each shaft allows for variations in distance between the transfer case and the axle units due to spring action. Two types of universal joints are used; the U-bolt type and the solid yoke type.

b. Tabulated Data.

(1) PROPELLER SHAFTS.

Make .... Spicer

Shaft diameter .... 1½ in.

Length (front) .... 21 11/16 in.

Length (rear) .... 20 1/32 in.

(2) FRONT PROPELLER SHAFT FORWARD UNIVERSAL JOINT.

Make .... Spicer

Type .... U-bolt and solid yoke

Model 1268

Bearings .... Needle roller

(3) FRONT PROPELLER SHAFT REAR UNIVERSAL JOINT.

Make .... Spicer

Type .... U-bolt and solid yoke

Model 1261

Bearings .... Needle roller

(4) REAR PROPELLER SHAFT FORWARD UNIVERSAL JOINT.

Make .... Spicer

Type .... Solid yoke slip joint

Model 1261

Bearings .... Needle roller

(5) REAR PROPELLER SHAFT REAR UNIVERSAL JOINT.

Make .... Spicer

Type .... U-bolt and solid yoke

Model 1268

Bearings .... Needle roller

17. REMOVAL.

a. **Front Propeller Shaft** (fig. 33). Bend the ears of the lock plates off the U-bolt nuts. Remove the two nuts from each of the two U-bolts at the front axle and at the transfer case. Remove the U-bolts from the propeller shaft. Take care to hold the bearing races in place on the universal joint to avoid losing the rollers.

b. **Rear Propeller Shaft** (fig. 34). The rear propeller shaft is similar to the front propeller shaft with the exception of the solid yoke type connection at the transfer case. Remove the nuts from the U-bolts at the rear axle end. Remove the U-bolts. Slide the universal joint out of the universal joint rear flange. Care must be taken to hold the bearing races on the universal joint to avoid losing the rollers. Remove the four nuts that secure the universal joint flange yoke to the rear flange at the transfer case. Remove the rear propeller shaft from the vehicle.

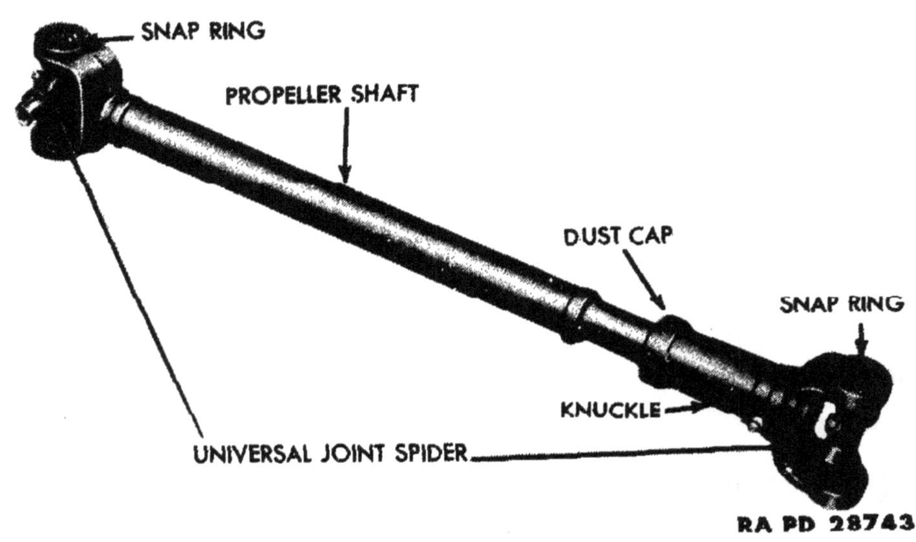

Figure 30 — Front Propeller Shaft

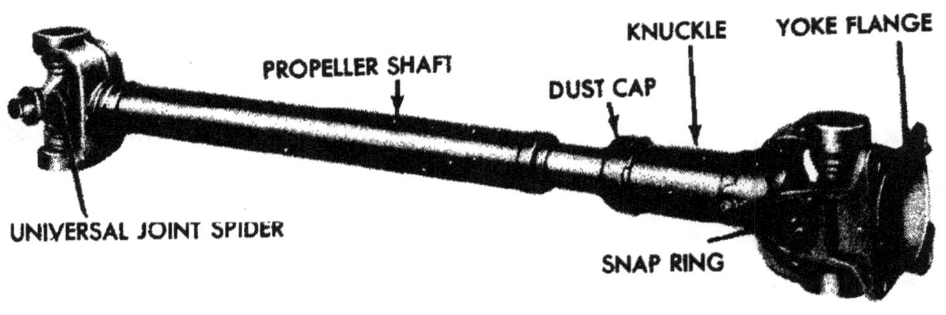

Figure 31 — Rear Propeller Shaft

18. DISASSEMBLY.

a. Front Propeller Shaft (fig. 30).

(1) REMOVE SNAP RINGS FROM YOKE (fig. 30). Place the propeller shaft in a vise. Remove the snap rings that secure the spider bearings in the yoke flange with a pair of pliers. If the snap ring does not snap out of the groove, tap the end of the bearing lightly. This will relieve the pressure against the snap ring.

(2) REMOVE SPIDER FROM YOKE (fig. 32). Drive lightly on the end of the spider bearing until the opposite bearing is pushed out of the yoke flange. Turn the assembly over in the vise and drive the first spider bearing back out of its lug by driving on the exposed end of the spider. Use a brass drift with a flat face about $\frac{1}{32}$ inch smaller

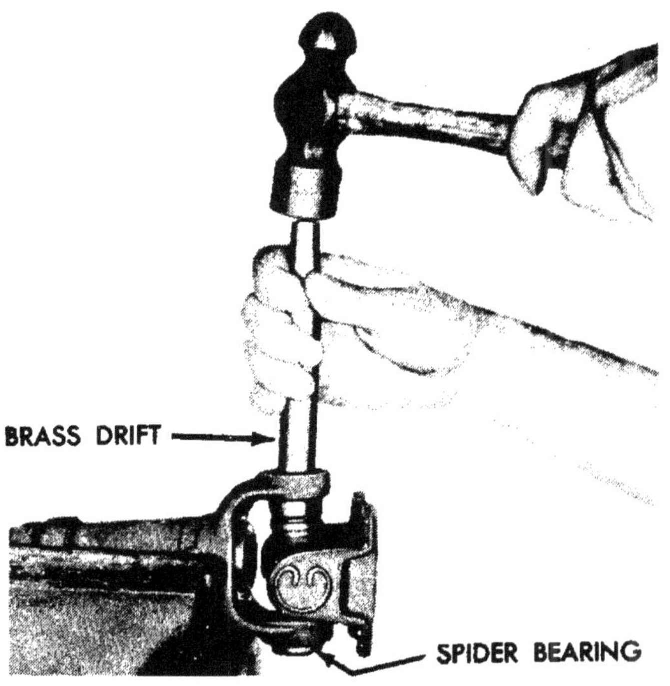

Figure 32 — Removing Spider Bearing

in diameter than the hole in the yoke, otherwise there is danger of damaging the spider bearing. Repeat this operation for the other two bearings, then lift out the spider, sliding to one side and tilting over the top of the yoke.

(3) REMOVE KNUCKLE FROM SHAFT (fig. 30). Bend the ears of the dust cap off the knuckle. Slide the knuckle off the drive shaft. Remove the split cork gasket from the bearing cap. Line up the slots in the dust cap with the splines on the drive shaft and remove the cap from the shaft.

b. Rear Propeller Shaft.

(1) REMOVE YOKE FLANGE (fig. 31). Place the propeller shaft in a vise. Remove the four snap rings that secure the spider bearings in the yoke flange and knuckle. Using a brass drift with a flat face about $\frac{1}{32}$ inch smaller than the hole in the yoke, drive lightly on the end of the bearing until the opposite bearing is out of the yoke flange. Turn the assembly over in the vise and drive the first bearing out of its lug by driving on the exposed end of the spider. Remove the yoke flange from the spider.

(2) REMOVE SPIDER AND KNUCKLE (fig. 32). Remove the spider from the knuckle (subpar. a (2), above). Remove the knuckle from the propeller shaft (subpar. a (3), above).

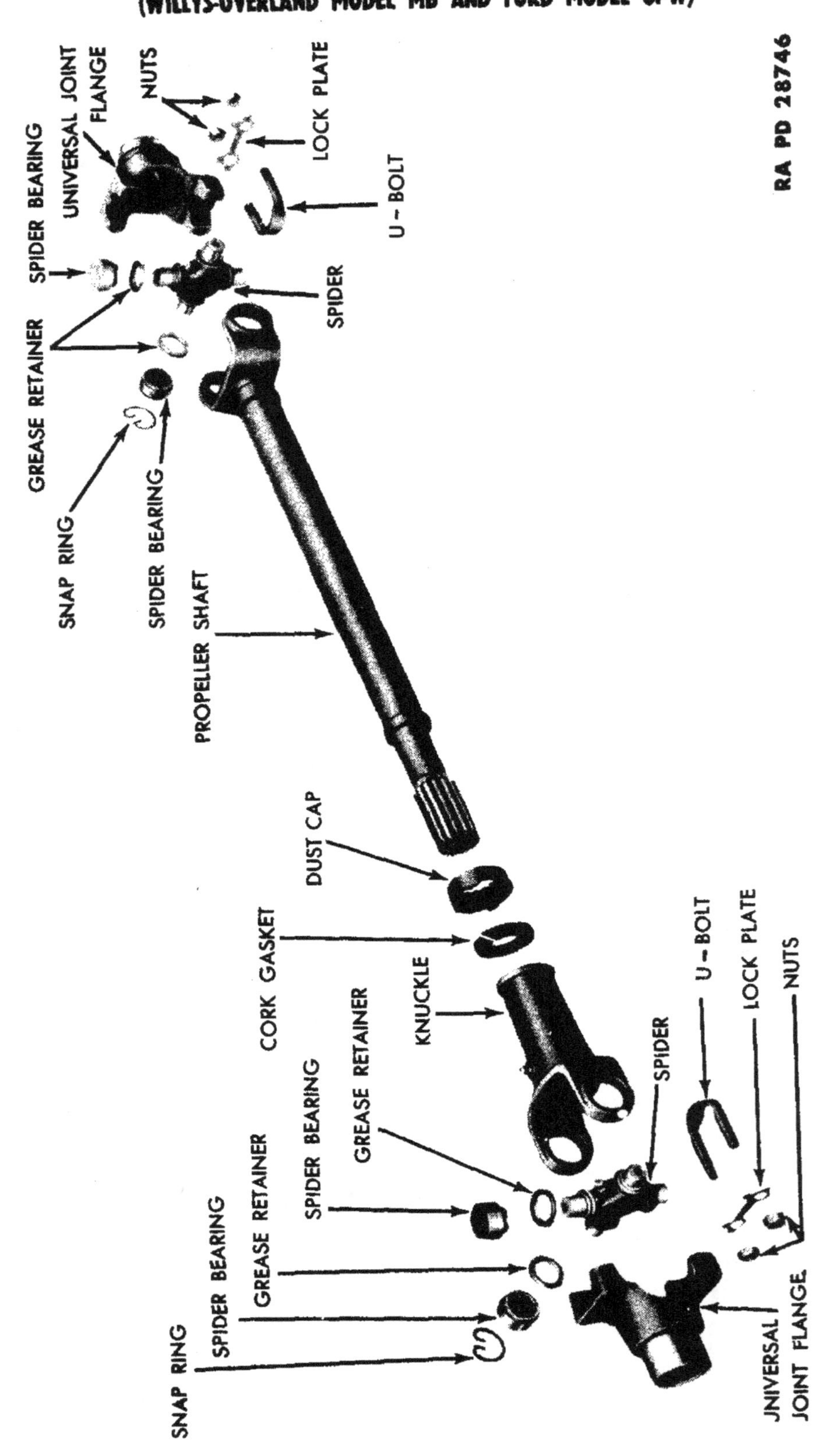

Figure 33 — Front Propeller Shaft — Exploded View

19. CLEANING, INSPECTION, AND REPAIR.

a. Clean all parts thoroughly with dry-cleaning solvent. Inspect the drive shafts for cracks, broken welds, scored spider bearing surfaces, or bent shafts. Parts with any of these faults must be replaced. Inspect the knuckle for worn splines, worn bearing surfaces and bearings and plugged lubricant fittings. Check the diameter of the machined surface of the spiders. If the diameter is less than 0.595 inch, replace the spider. Replace all grease seals regardless of their condition.

20. ASSEMBLY.

a. Front Propeller Shaft (fig. 33). Place the propeller shaft in a vise. Slide the dust cap on the drive shaft. Place a new cork gasket in the cap. Slide the knuckle on the shaft splines, being sure that the knuckle on the shaft is in the same angle as the yoke at the opposite end of the propeller shaft. Slide the dust cap on the shoulder of the knuckle and bend the ears of the cap over the shoulder of the knuckle.

b. Rear Propeller Shaft (fig. 34).

(1) INSTALL SPIDER IN YOKE FLANGE (fig. 34). Insert the spider into the yoke flange. Tap the spider bearing approximately 1/4 inch into the yoke flange, using a brass drift approximately 1/32 inch smaller than the hole in the yoke. Tap the other bearing into the opposite end of the yoke flange until the bearing is in line with the snap ring grooves. With a pair of pliers, install the snap rings on both ends of the yoke flange. Insert the flange assembly in the knuckle. Tap the bearing approximately 1/4 inch into the yoke. Place the other bearing into the opposite end of the yoke, and tap this bearing into the yoke until the bearing is in line with the snap ring groove. Install the snap rings on both ends of the yoke.

(2) INSTALL KNUCKLE AND SPIDERS (fig. 34). Install the knuckle on the propeller shaft (subpar. a (1), above).

21. INSTALLATION.

a. Rear Propeller Shaft. Place the propeller shaft with the yoke flange end toward the transfer case (fig. 6). Install the four nuts that secure the yoke flange to the transfer case. Insert the two spider bearings on the spider at the rear axle end. Place the spider in the universal joint rear flange. Install the two U-bolts that secure the propeller shaft to the rear axle flange. Lubricate the propeller shaft with specified lubricant.

b. Front Propeller Shaft. Place the propeller shaft with the knuckle end at the transfer case. Insert the bearings on the spider

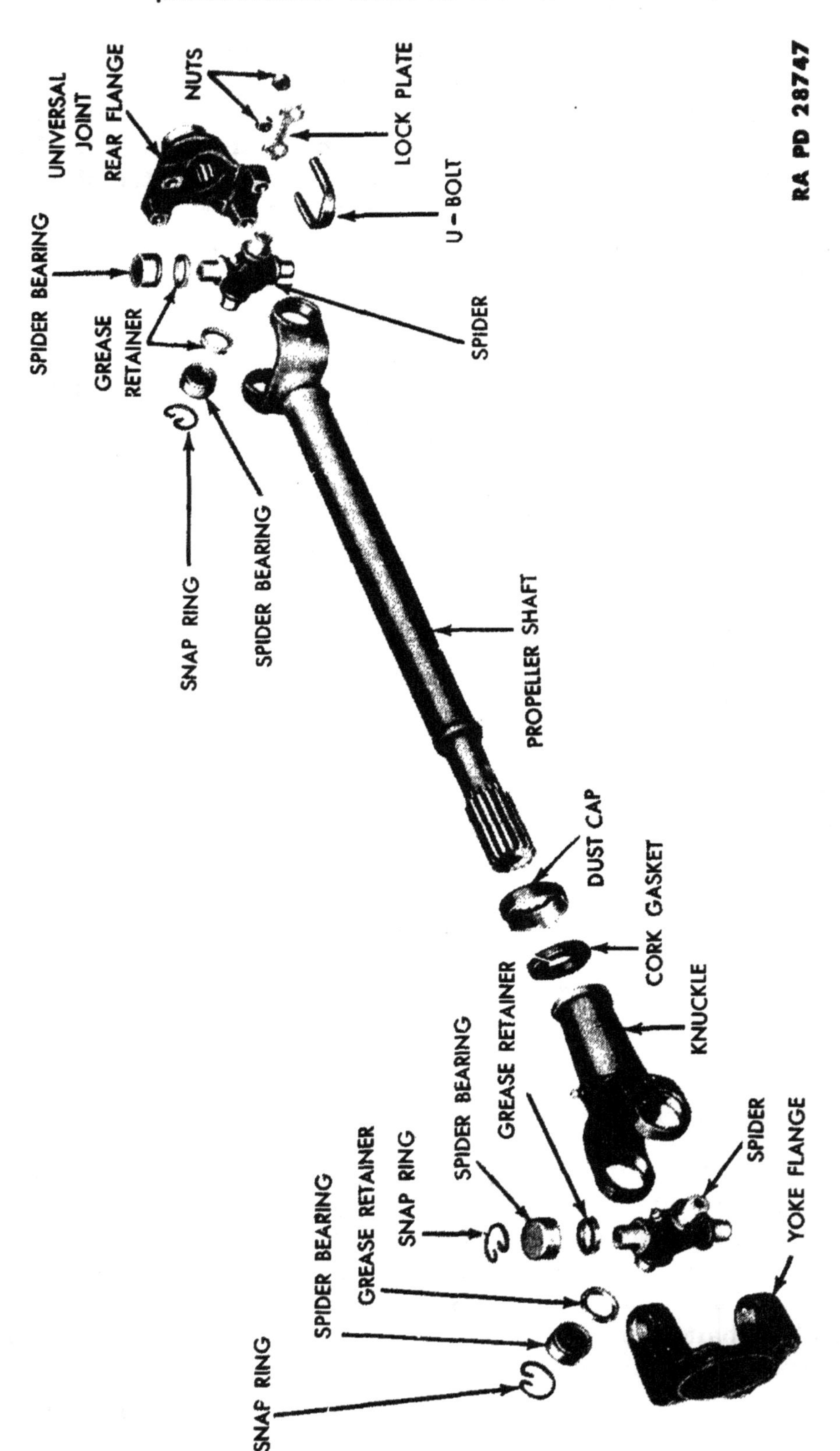

Figure 34 — Rear Propeller Shaft — Exploded View

and place the propeller shaft in the universal joint flange on the transfer case. Install the two U-bolts that secure the propeller shaft to the transfer case. Insert the two spider bearings on the spider at the front axle end. Place the propeller shaft in the front axle flange. Install the two U-bolts that secure the propeller shaft to the universal joint flange. Lubricate the propeller shaft with specified lubricant.

Section V

FRONT AXLE

22. DESCRIPTION AND DATA.

a. **Description** (fig. 2). The front axle assembly is a front wheel driving unit, with specially designed spindle housings, and has a conventional type differential with hypoid drive gears. The differential parts are interchangeable with those of the rear axle. The axle shafts are of the full-floating type. The differential is mounted in the housing similar to the rear axle, except that the drive pinion shaft is toward the rear instead of the front and to the right of the center of the axle. Three types of axle shafts and universal joints have been used (Rzeppa, Bendix, and Tracta). The vehicles using the different types of shafts are identified by an identification tag attached to the spindle housing (fig. 35).

b. **Data.**

(1) FRONT AXLE.

Make .... Spicer

Drive .... Through springs

Type .... Full-floating

(2) DIFFERENTIAL.

Drive .... Hypoid

Gear ratio .... 4.88 to 1

Bearings .... Timken roller 2

Adjustment .... Shims

Gears (pinion) 2

(3) OIL CAPACITY .... 2½ pt

Figure 35 — Front Axle Assembly in Vehicle

A—SPRING SHACKLE
B—SHOCK ABSORBER
C—TIE ROD
D—BREATHER CAP
E—TIE ROD ENDS
F—SPRING SHACKLE
G—TIE ROD
H—LEFT FRONT SPRING
J—SHOCK ABSORBER
K—TIE ROD CLAMP
L—TORQUE REACTION SPRING
M—DRAG LINK
N—DRAG LINK PLUG
O—PIVOT ARM
P—DRAIN PLUG
Q—FRONT PROPELLER SHAFT
R—SPRING SEAT PLATE
S—TIE ROD CLAMP
T—TIE ROD ENDS
U—RIGHT FRONT SPRING
V—TIE ROD CLAMPS
W—AXLE SHAFT IDENTIFICATION TAG

RA PD 329205-B

Legend for Figure 35 — Front Axle Assembly in Vehicle

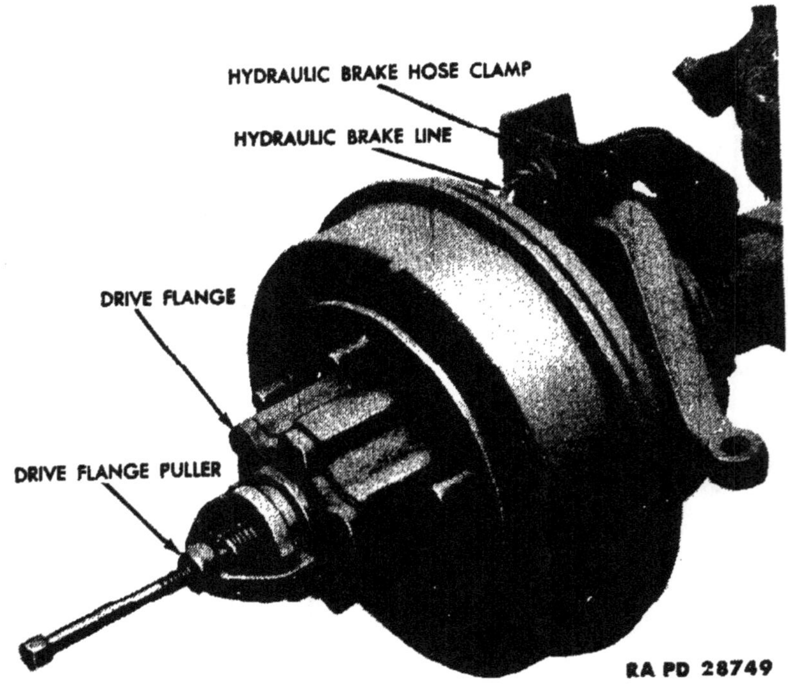

Figure 36 — Removing Drive Flange With Puller Similar to Puller 41-P-2912

23. REMOVAL.

a. **Preliminary Work.** Remove the drain plug at the differential housing and drain the oil. Raise the vehicle until the weight is off the front springs.

b. **Disconnect Shock Absorbers and Drag Link** (fig. 35). Remove the cotter pin and flat washer that secure the shock absorber to the spring seat plate at both front shock absorbers. Remove the drag link plug at the pivot arm. Remove the drag link from the pivot arm.

c. **Disconnect Front Propeller Shaft and Spring U-bolts** (fig. 35). Disconnect the front propeller shaft at the front anxle (par. 17 a). Remove the four nuts from the two U-bolts that secure the spring seat plate. Remove the spring seat plate and U-bolts. Remove the four nuts from the U-bolts at the torque reaction spring. Remove the two U-bolts.

d. **Disconnect Spring Shackles** (fig. 35). Remove the lower spring shackle bushing at the forward end of the front springs. Pull both springs out of the spring shackles and drop the forward end of the springs to the floor. Roll the front axle assembly from the vehicle.

Figure 37 — Removing Bearing Lock Nut With Wrench 41-W-3825-200

24. DISASSEMBLY.

a. **Remove Wheels.** Place the front axle assembly on two blocks. Remove the five nuts that secure the wheels to the brake drum. Remove the wheels.

b. **Remove Axle Shaft Assembly.** Using a screwdriver, pry the hub cap off the drive flange. Remove the cotter pin and castellated nut from the axle shaft. Remove the six cap screws that secure the drive flange to the hub. Install the puller 41-P-2912 or similar on the drive flange and remove the drive flange (fig. 36). Bend the ear of the lock washer off the bearing lock nut. Remove the bearing lock nut, lock washer, and bearing adjustment nut, using the wheel bearing nut wrench 41-W-3825-200 furnished with the vehicle (fig. 37). Slide the brake drum and hub assembly, including the wheel bearings, off the spindle. Disconnect the hydraulic brake line at the brake hose guard (fig. 36). Remove the six cap screws that secure the brake plate to the spindle housing. Remove the brake plate from the spindle. Slide the spindle off the axle shaft. The axle shaft can now be removed from the housing. If equipped with a Tracta universal joint axle shaft, see subparagraph c, below. Use the same procedure to disassemble the other end of the front axle shaft.

c. **Axle Shaft Disassembly.** Three types of axle shaft universal joints, as shown in figures 38, 42, and 44, are used in the front axle.

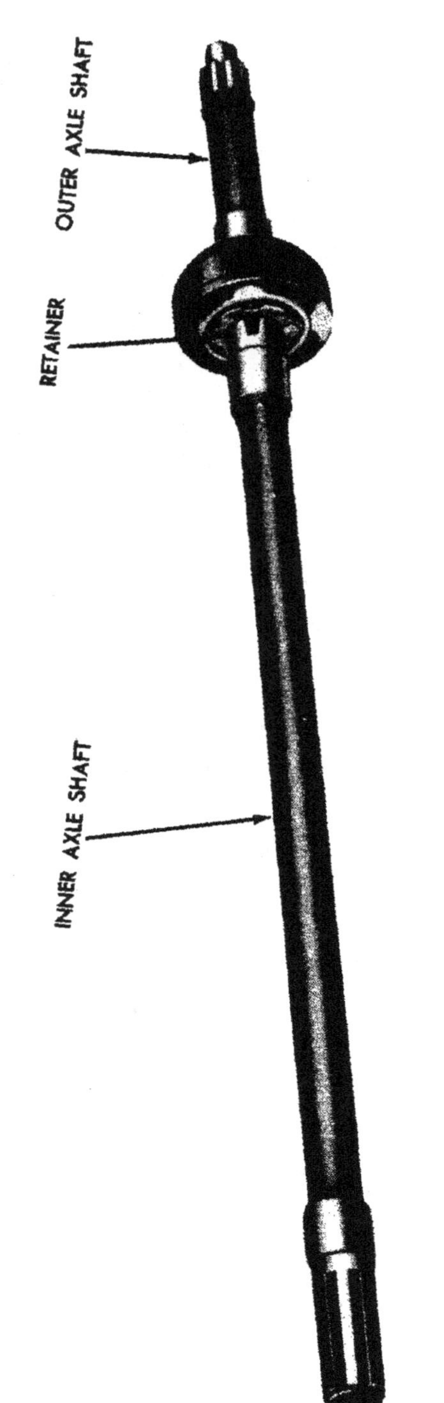

RA PD 28751

Figure 38 — Front Axle Shaft (Rzeppa Joint)

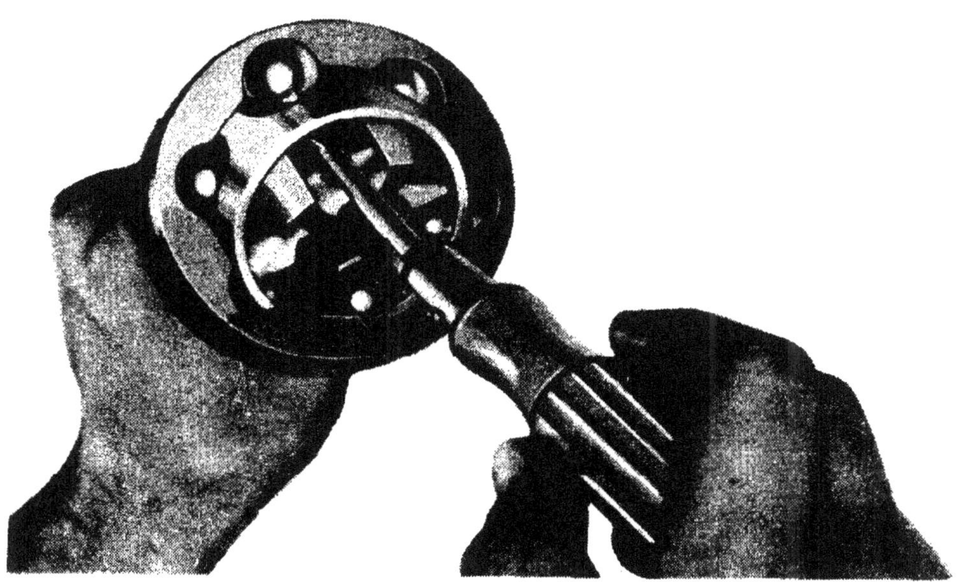

RA PD 28752

Figure 39 — Removing Balls From Cage

Disassembly procedures for each are given in steps (1), (2), and (3), below.

(1) RZEPPA UNIVERSAL JOINT.

(a) Remove Inner Axle Shaft (fig. 59). Remove the three flat head screws that secure the retainer to the inner ball race. Slide the inner axle shaft out of the universal joint. Remove the pilot pin from the outer axle shaft. If the pilot pin does not drop out of the outer axle shaft, hold the shaft upside down and tap the shaft on a piece of wood.

(b) Remove Balls From Cage (fig. 39). Tilt the cage in the axle shaft cup until the opposite side of the cage is out of the housing. It may be necessary to use a brass drift and hammer to tilt the cage. Use a screwdriver to pry the steel ball out of the cage. Repeat this operation until all the balls are removed.

(c) Remove Cage and Inner Race From Axle Shaft (fig. 40). Turn the cage in the axle shaft cup in line with the shaft and with the two larger elongated holes between two bosses in the shaft. Lift the cage and inner race from the axle shaft cup.

(d) Remove Inner Race From Cage (fig. 41). Turn the inner race in the cage so that one of the bosses on the inner race can be dropped into one of the two elongated holes in the cage. Remove the inner race from the cage.

(2) BENDIX UNIVERSAL JOINT (figs. 42 and 43). Place the axle

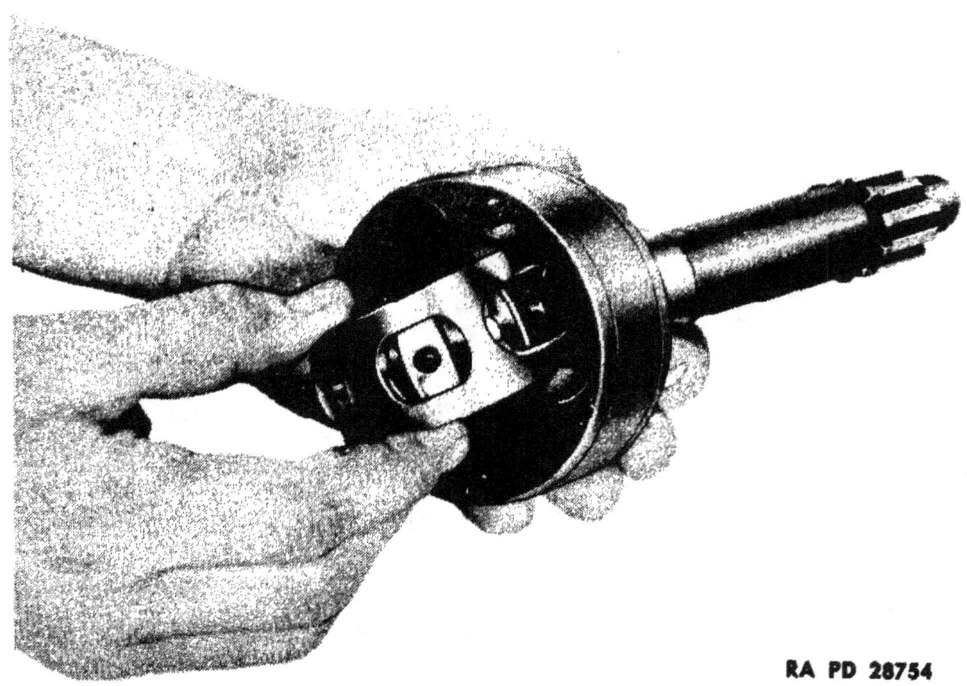

RA PD 28754

Figure 40 — Removing Cage and Inner Race From Axle Shaft (Rzeppa Joint)

RA PD 28753

Figure 41 — Removing Inner Race From Cage (Rzeppa Joint)

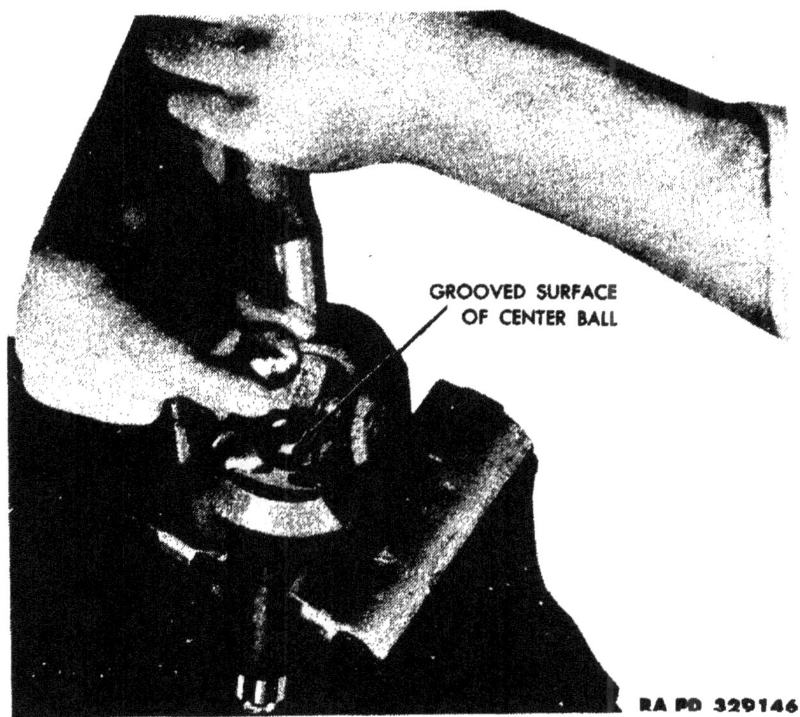

Figure 42 — Front Axle Shaft (Bendix Type)

shaft in a vise and with a long nosed drift remove the groove pin from the universal joint knuckle. Remove the axle shaft from the vise. Tap the knuckle end of the axle shaft on a wood block until the center ball pin drops in the groove pin hole. Place the axle shaft with the knuckle end (short end) in a vise. Bend the axle shaft so that the center ball can be rotated until the grooved surface of the center ball is facing the first ball that is to be removed. Holding the axle shaft in a bent position, raise the shaft until the first ball to be removed slides into the groove of the center ball, and remove the ball. Remove the axle shaft from the knuckle. The three remaining balls will drop out of the knuckle.

(3) TRACTA UNIVERSAL JOINT (fig. 45). Remove the outer portion of the axle shaft and the outer portion of the universal joint from the axle housing. Pull the inner portion of the axle shaft and the inner portion of the universal joint out of the housing.

d. Remove Spindle Housing (fig. 46). Remove the castellated nut that secures the tie rod ends to the two spindle arms. Remove the two castellated nuts that secure the two tie rod ends to the steering pivot arm and remove the two tie rods. Remove the hydraulic brake hose clamp from the hydraulic brake line at the brake hose guard. Remove the four nuts that secure the brake hose guard and spindle arm to the spindle housing. Remove the spindle arm and

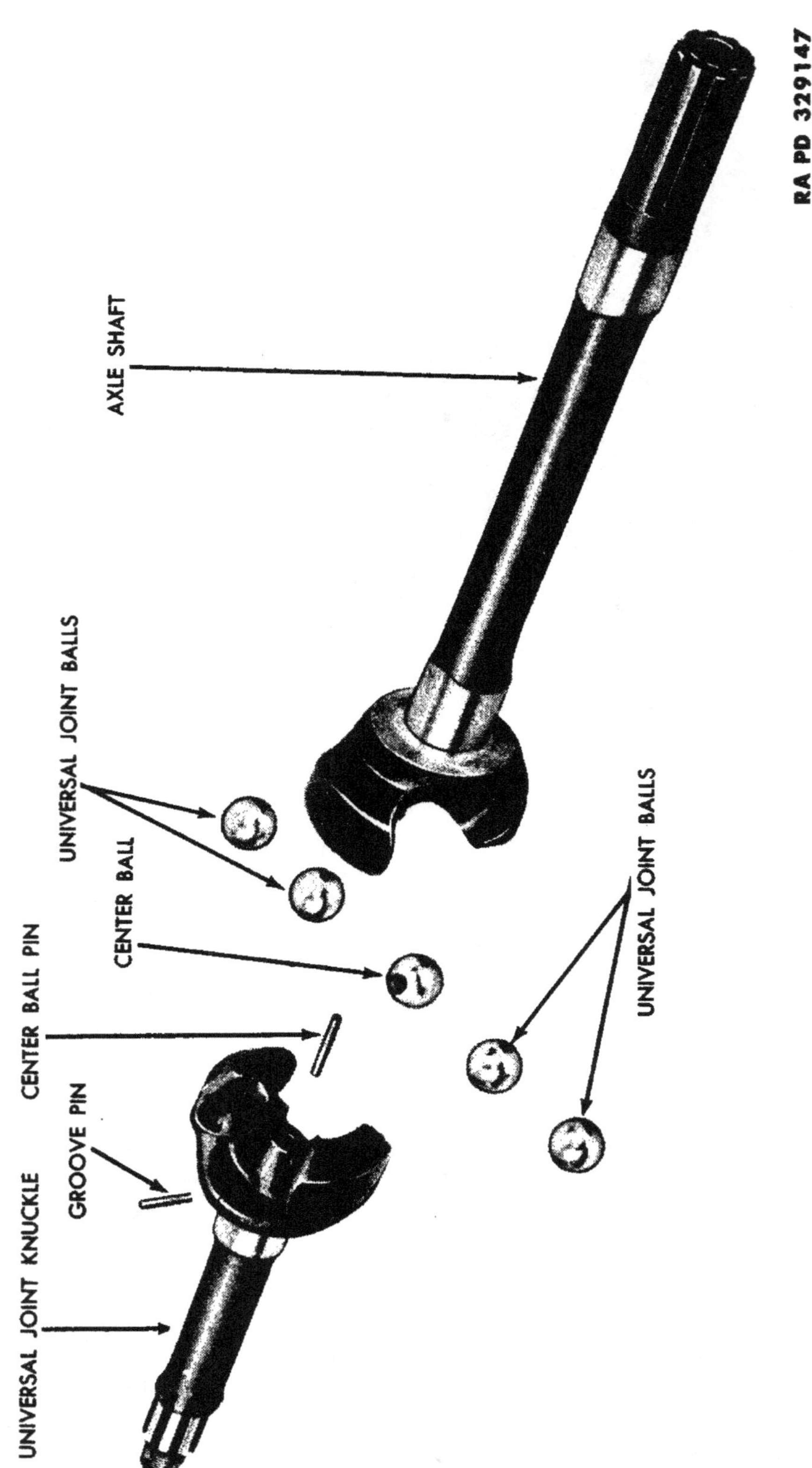

Figure 43 — Front Axle Shaft — Exploded View (Bendix Type)

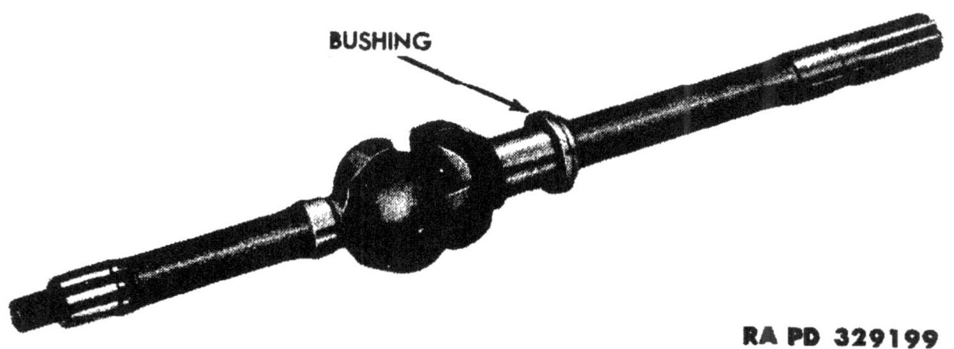

Figure 44 — Front Axle Shaft (Tracta Type)

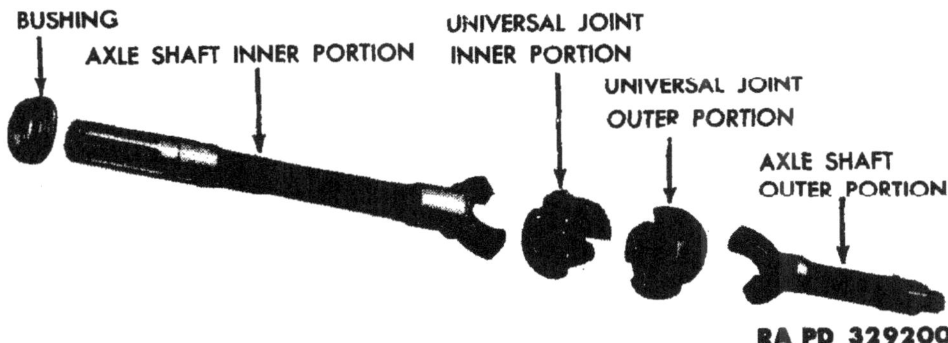

Figure 45 — Front Axle Shaft — Exploded View (Tracta Type)

shims from the spindle housing. Remove the four cap screws that secure the lower bearing cap to the spindle housing. Remove the bearing cap and shims. Remove the eight cap screws that secure the spindle housing oil seal retainer to the spindle housing. Remove the spindle housing from the axle housing. Use the same procedure for disassembling the other spindle housing.

e. **Remove Differential** (fig. 47). Remove the ten cap screws that secure the differential cover to the differential housing. Remove the differential cover and gasket. Remove the two cap screws from the bearing cap at each end of the differential gears and remove the caps. Remove the differential gear assembly from the housing, using a pry bar, if necessary. Reinstall the bearing caps in the housing, noting the markings (fig. 47) to assure their being installed in their correct location.

f. **Disassemble Differential.**

(1) REMOVE DIFFERENTIAL PINION GEARS AND AXLE SHAFT GEARS (fig. 48). Place the differential assembly in a vise equipped with brass jaws. With a long nosed drift, drive the differential

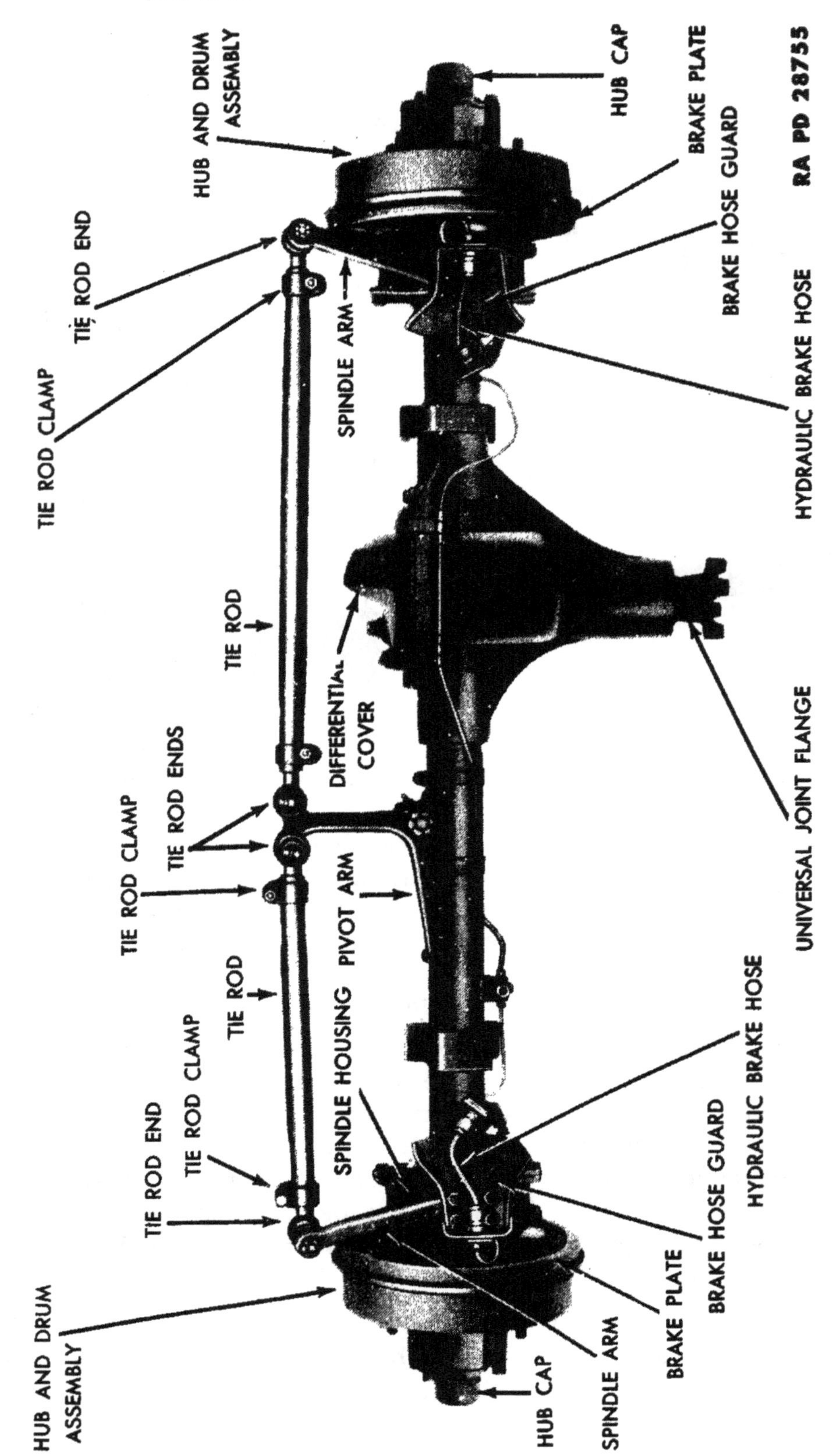

Figure 46 — Front Axle Assembly

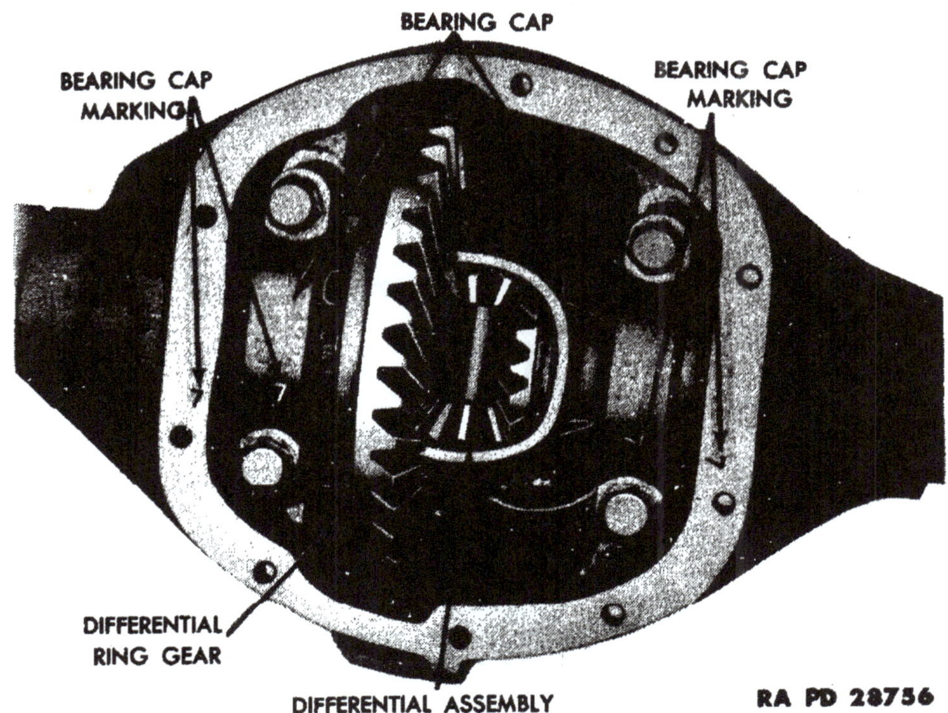

Figure 47 — Differential Assembly

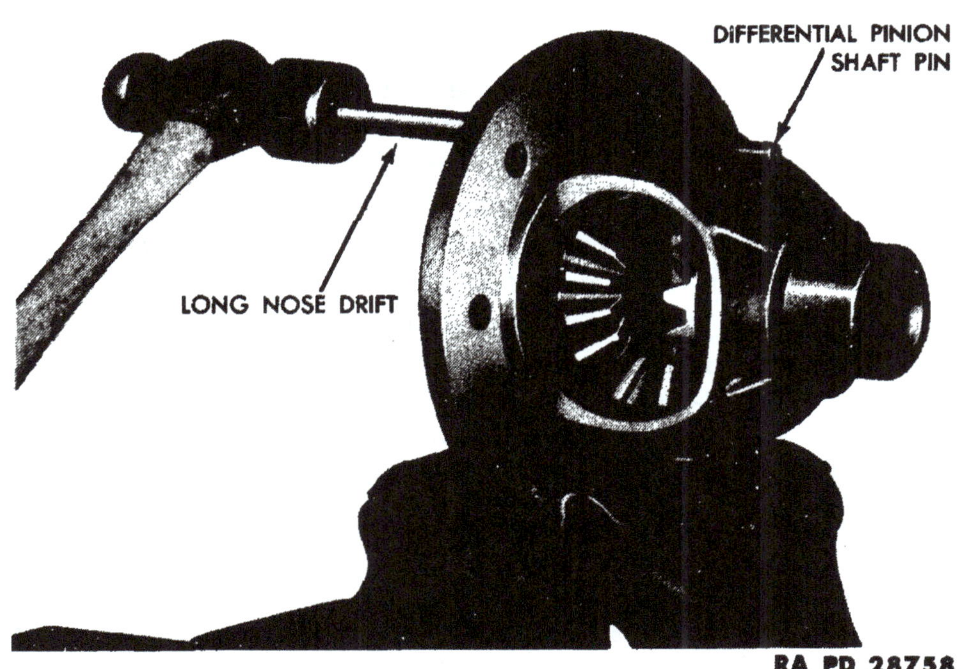

Figure 48 — Removing Pinion Shaft Lock Pin

RA PD 28757

Figure 49 — Removing Bearings From Differential Case With Special Tool 41-R-2378-30

pinion shaft tapered pin out of the differential gear case. Tap the differential pinion shaft from the case with a brass drift and hammer. Remove the two differential pinion gears and thrust washers and the two axle shaft gears and thrust washers from the case.

(2) REMOVE RING GEAR FROM CASE (fig. 47). Bend the ears of the lock plates off the cap screws. Remove the cap screws that secure the ring gear to the case, and remove the ring gear.

(3) REMOVE ROLLER BEARING FROM DIFFERENTIAL CASE (fig. 49). Place the differential case in a vise. Install the bearing remover 41-R-2378-30 to the roller bearing. Remove the roller bearing from each end of the differential case. Remove the shims, noting the thickness of the shims removed from each end.

g. **Remove Drive Pinion.** Remove the nut and flat washer that secure the universal joint flange to the drive pinion. Install the puller 41-P-2905-60 to the universal joint flange (fig. 50) and remove the flange. Using a brass drift and hammer, drive the drive pinion out of the axle housing (fig. 51). Remove the shims and spacer from the drive pinion, noting the thickness of the shims removed from the pinion.

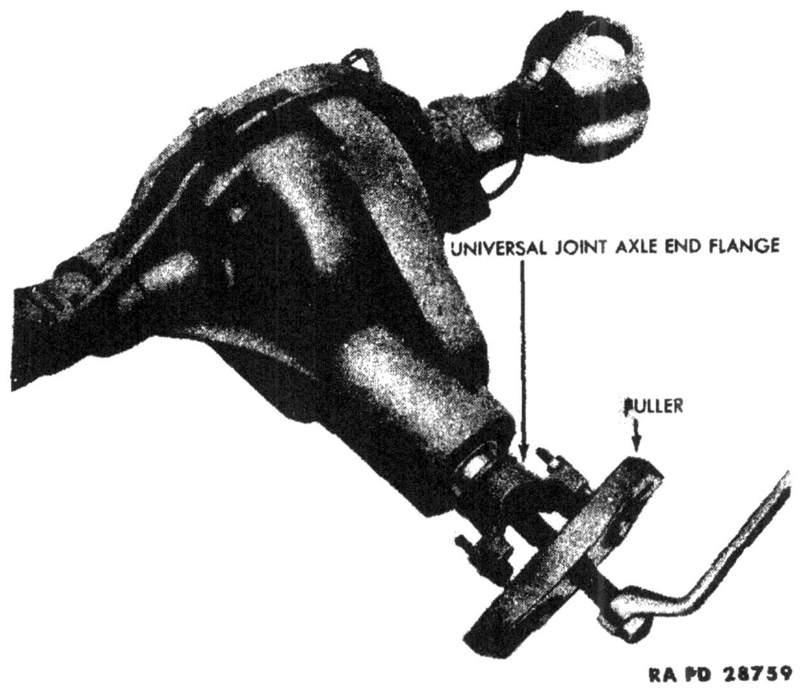

Figure 50 — Removing Universal Joint Axle End Flange With Puller 41-P-2905-60

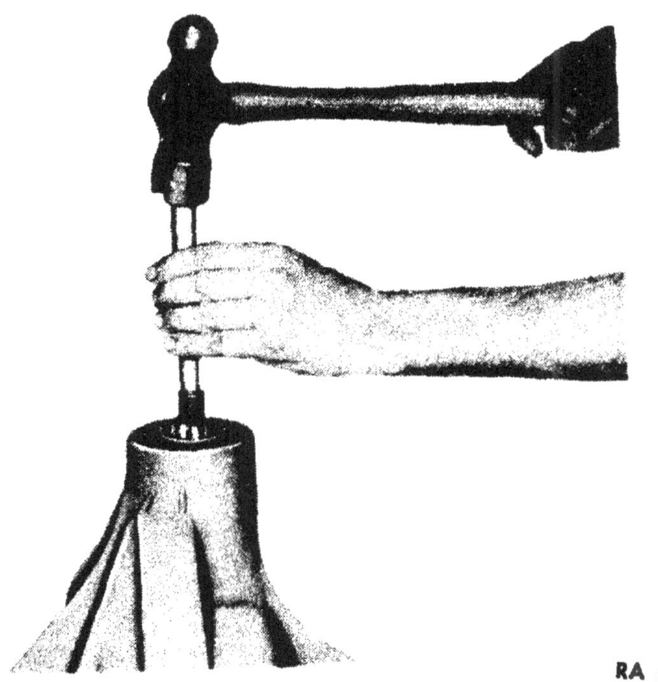

Figure 51 — Removing Drive Pinion

RA PD 28761

Figure 52 — Installing Pinion Outer Bearing Cup

25. CLEANING, INSPECTION, AND REPAIR.

a. Cleaning. Clean all parts in dry-cleaning solvent. Rotate the bearings while immersed in the dry-cleaning solvent until all trace of lubricant has been removed. Oil the bearings to prevent corrosion of the highly polished surface unless they are to be used immediately.

b. Inspection and Repair.

(1) AXLE HOUSING (fig. 53).

(a) Inspection. Replace the axle housing if it is bent or has any broken welds or cracks. Drive pinion bearing cups that are pitted, corroded or discolored due to overheating must be replaced (step *(c)*, below). Spindle housing bearing cups that are pitted or corroded must be replaced (step *(d)*, below). Replace the oil seals in the axle housing regardless of their condition (step *(e)*, below). Replace the differential cover, if cracked or if it has damaged threads in the filler plug hole. Check the cover for missing or damaged breather cap. Check the steering pivot arm shaft. If the diameter is less than 0.747 inch, replace the pivot shaft (step *(b)*, below). If the front axle is equipped with a Tracta type axle shaft, measure the

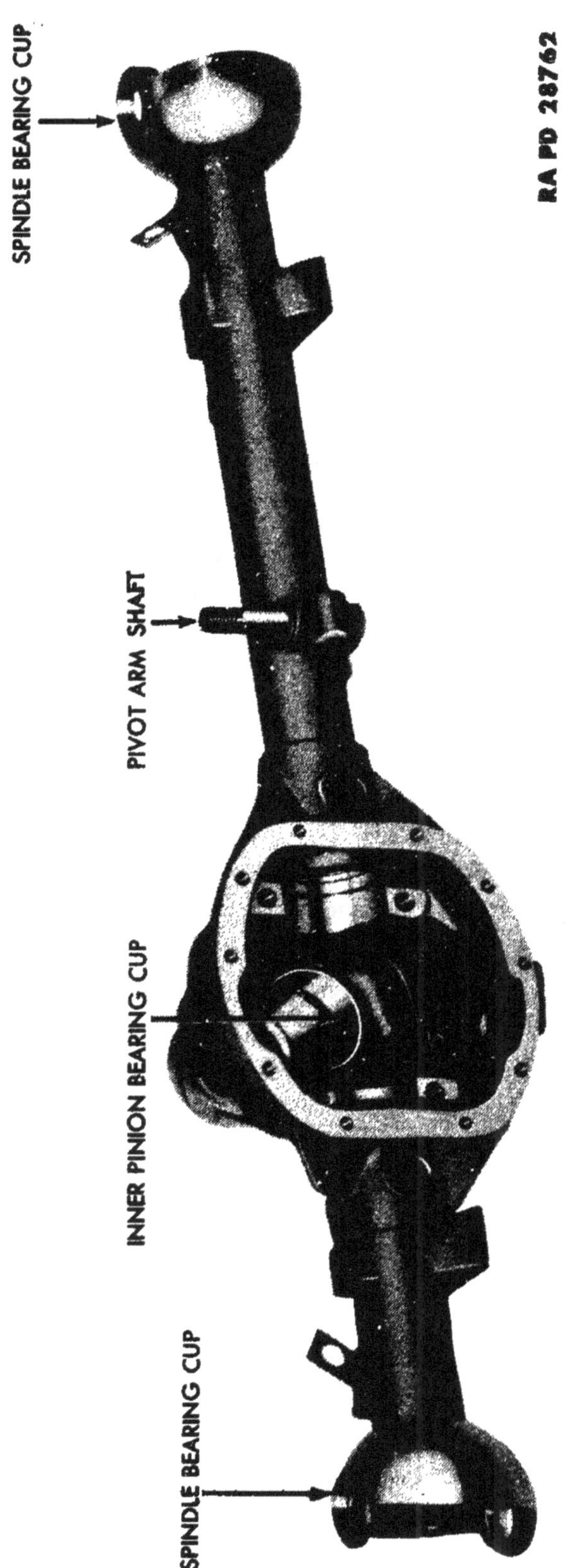

Figure 53 — Front Axle Housing

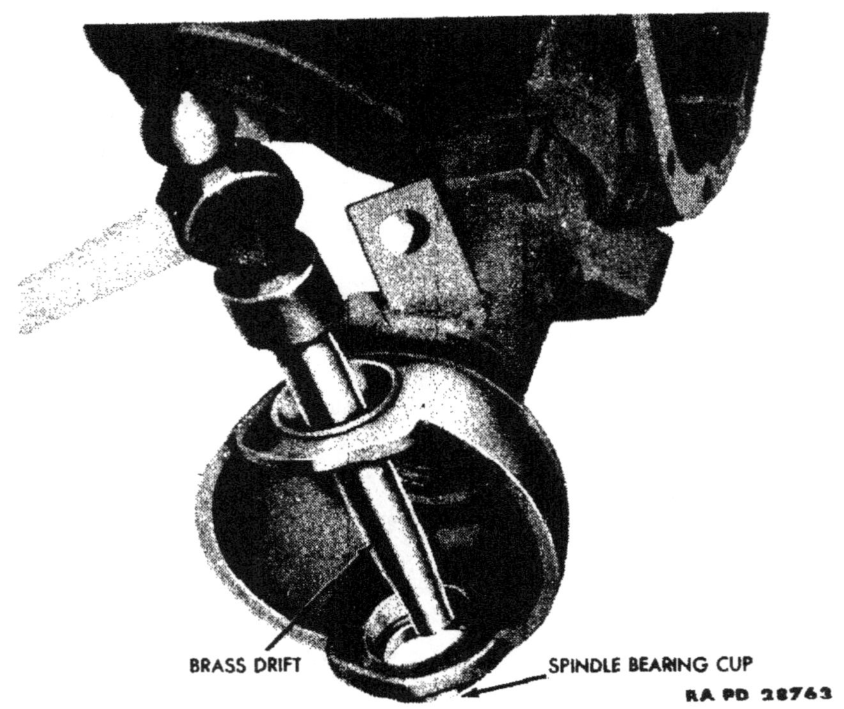

Figure 54 — Removing Spindle Bearing Cup From Axle Housing

inside diameter of the housing at each end of the axle housing. If the bushing is worn to more than 1.285 inch, replace the bushing (step *(f)*, below).

(b) Pivot Arm Shaft Replacement (fig. 53). With a long nosed drift, drive out the dowel that secures the pivot arm shaft to the axle housing. Tap the shaft out of the housing. To install a new pivot arm shaft, insert it in the bracket on the housing with the dowel slot in line with the dowel hole. Drive dowel in place.

(c) Drive Pinion Bearing Cup Replacement. Remove the inner and outer bearing cups, using a standard puller, noting the thickness of the shims when removing the inner bearing cup. To install new bearing cups, use a brass drift and hammer. Place the original thickness of shims behind the inner bearing cup and tap the bearing cups lightly around the entire circumference until flush with the shoulder in the axle housing (fig. 52).

(d) Spindle Housing Bearing Cup Replacement. Working through one of the bearing cups, tap the opposite bearing cup out of the axle housing, using a brass drift and hammer (fig. 54). To install new bearing cups, place the bearing cup in position and tap the cup lightly until it is flush with the shoulder in the axle housing.

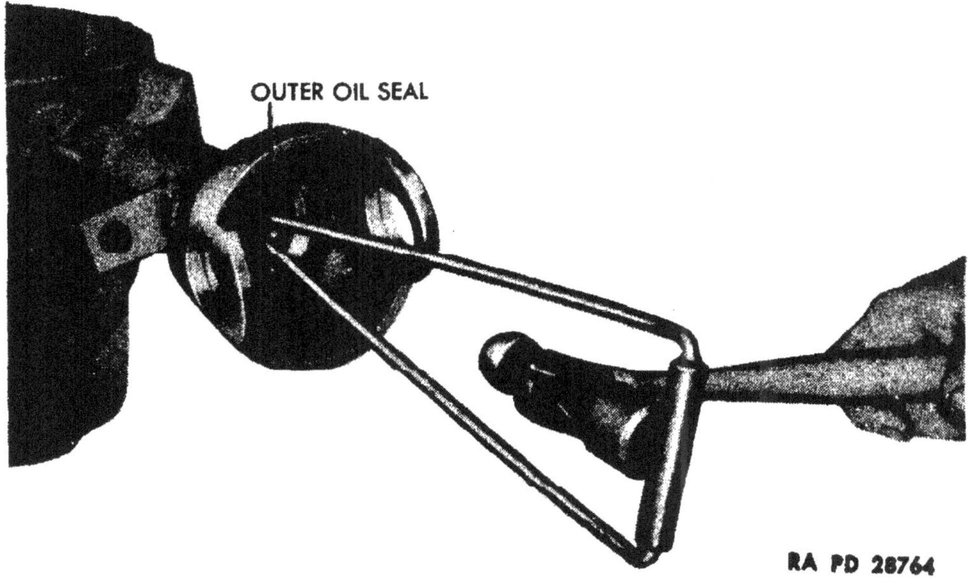

Figure 55 — Removing Oil Seal From Outer End of Axle Housing With Remover 41-R-2384-38

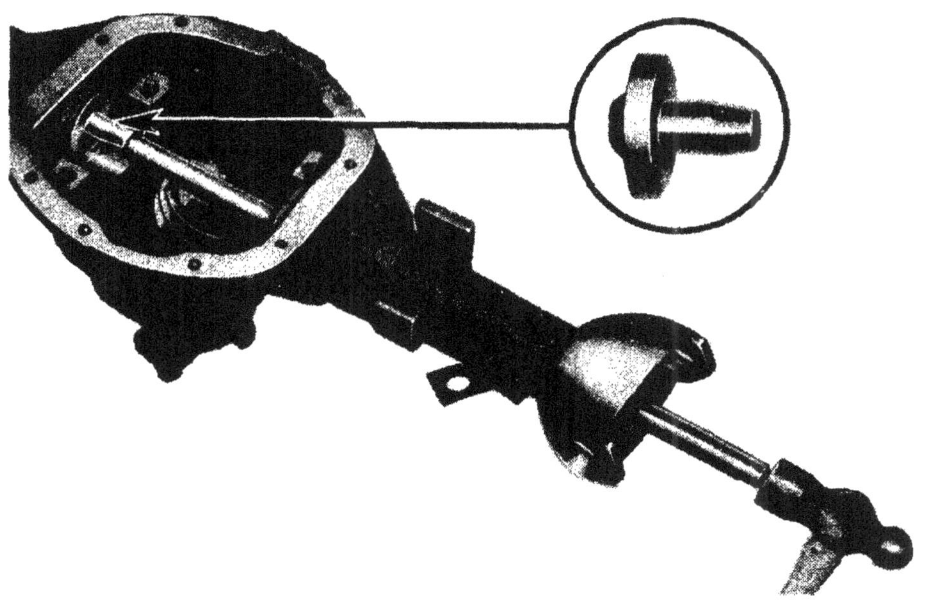

Figure 56 — Installing Oil Seal, With Replacer 41-R-2391-20

(e) *Oil Seal Replacement* (fig. 55). To remove the outer axle shaft oil seal, remove the oil seal retainer. Use a screwdriver to pry the retainer out of the housing. Use the oil seal remover 41-R-2384-38 to remove the inner and outer oil seals (figs. 55 and 80). To install the inner and outer oil seals, use the oil seal replacer

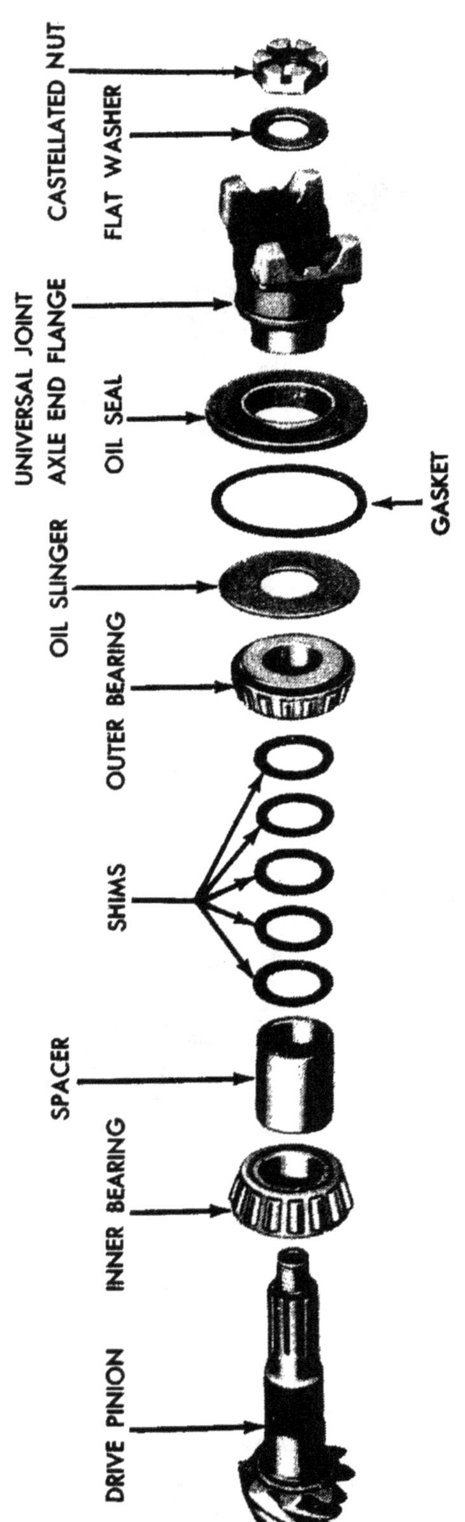

Figure 57 — Drive Pinion Assembly — Exploded View

41-R-2391-20 and tap the oil seals in the inner and outer ends of the axle housing (fig. 56). Using a brass hammer, tap the oil seal retainer in the outer end of the axle shaft housing.

(f) Axle Housing Bushing Replacement (For Tracta Type Axle Shafts Only). Remove the bushing from the outer end of the axle housing, using a standard puller. To install a new bushing, place the bushing in position in the axle housing and using a suitable driver, drive the bushing in the housing until it is flush with the shoulder in the axle housing.

(2) DRIVE PINION ASSEMBLY (fig. 57). Roller bearings that are pitted, corroded or discolored due to overheating must be replaced. Replace the drive pinion if it has worn or broken teeth. The differential ring gear and the drive pinion assembly are furnished only in matched sets and if either is found damaged, both must be replaced. Small nicks can be removed from the pinion gear with a fine stone.

(3) DIFFERENTIAL ASSEMBLY (fig. 58). Replace any gear that is excessively worn or has any broken teeth. The differential ring gear and the drive pinion assembly are furnished only in matched sets and if either is found damaged, both must be replaced. Replace the differential pinion gears, if the inside diameter is worn to more than 0.627 inch. Replace the differential pinion shaft if the diameter is less than 0.623 inch. Replace the axle shaft gears if the outside diameter of the hub is worn to less than 1.498 inches. Replace the differential pinion gear and axle shaft gear thrust washers if the thickness is worn to less than 0.32 inch. Roller bearings and races that are pitted, corroded or discolored due to overheating must be replaced. All shims that were damaged during the disassembly must be replaced.

(4) AXLE SHAFTS. Three different types of axle shaft universal joints as shown in figures 38, 42, and 44 are used in front axles. Inspection of each of these three types is covered in steps *(a)*, *(b)*, and *(c)*, below.

(a) Rzeppa Universal Joint (fig. 59). Replace the inner axle shaft if it is bent or has worn splines. Using a new axle shaft gear as a gage, slip it on the inner axle shaft and check the backlash. If the backlash is more than 0.005 inch, replace the axle shaft. Replace the outer axle shaft if it has worn splines or nicked ball bearing surfaces. Replace the inner race if it is found to be excessively worn. Small nicks or scratches can be removed with a fine stone. Replace steel balls that have flat spots. Replace the cage if it is cracked.

(b) Bendix Universal Joint (fig. 43). Replace the inner axle shaft if it is bent or has worn splines or worn universal joint ball surface. Replace the universal joint knuckle if it has worn splines or

855473 O - 49 - 5

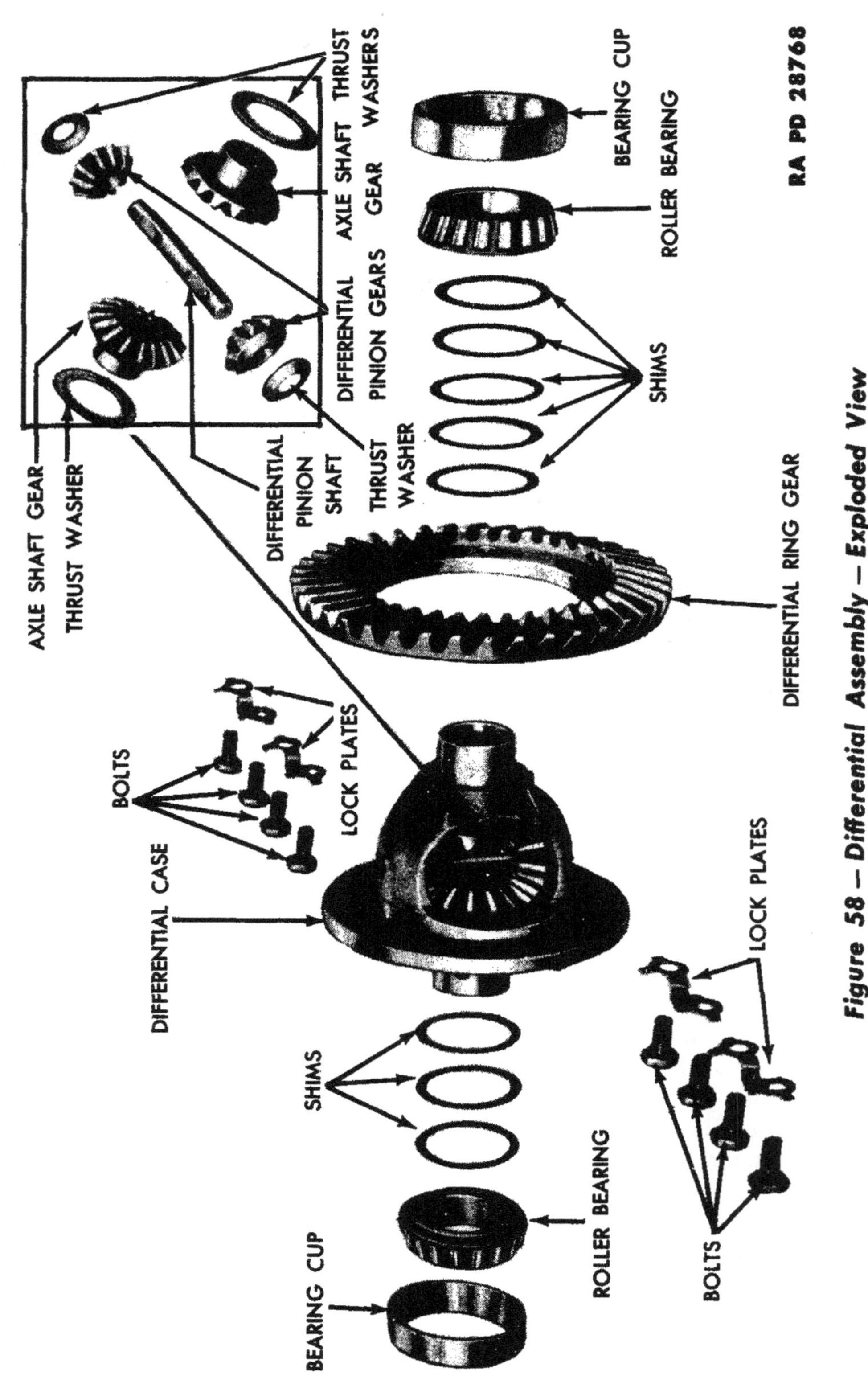

Figure 58 — Differential Assembly — Exploded View

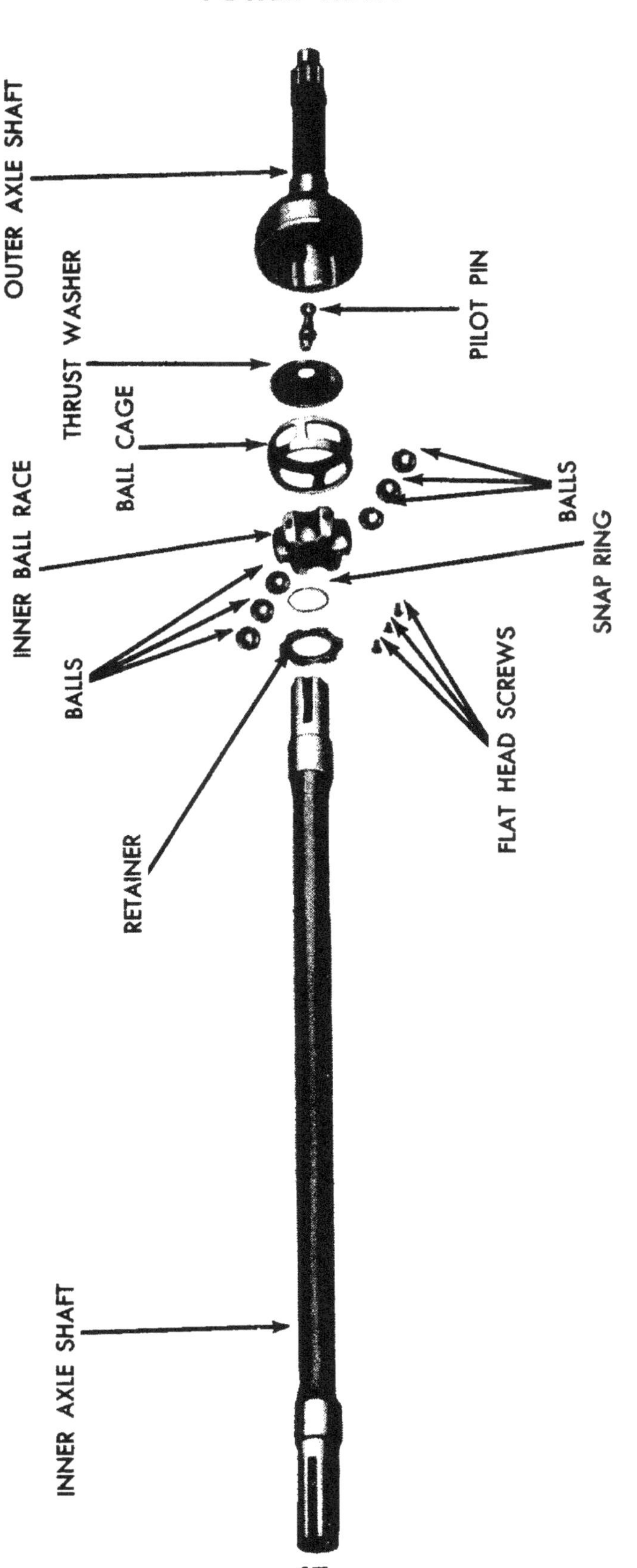

Figure 59 — Axle Shaft — Exploded View (Rzeppa Type)

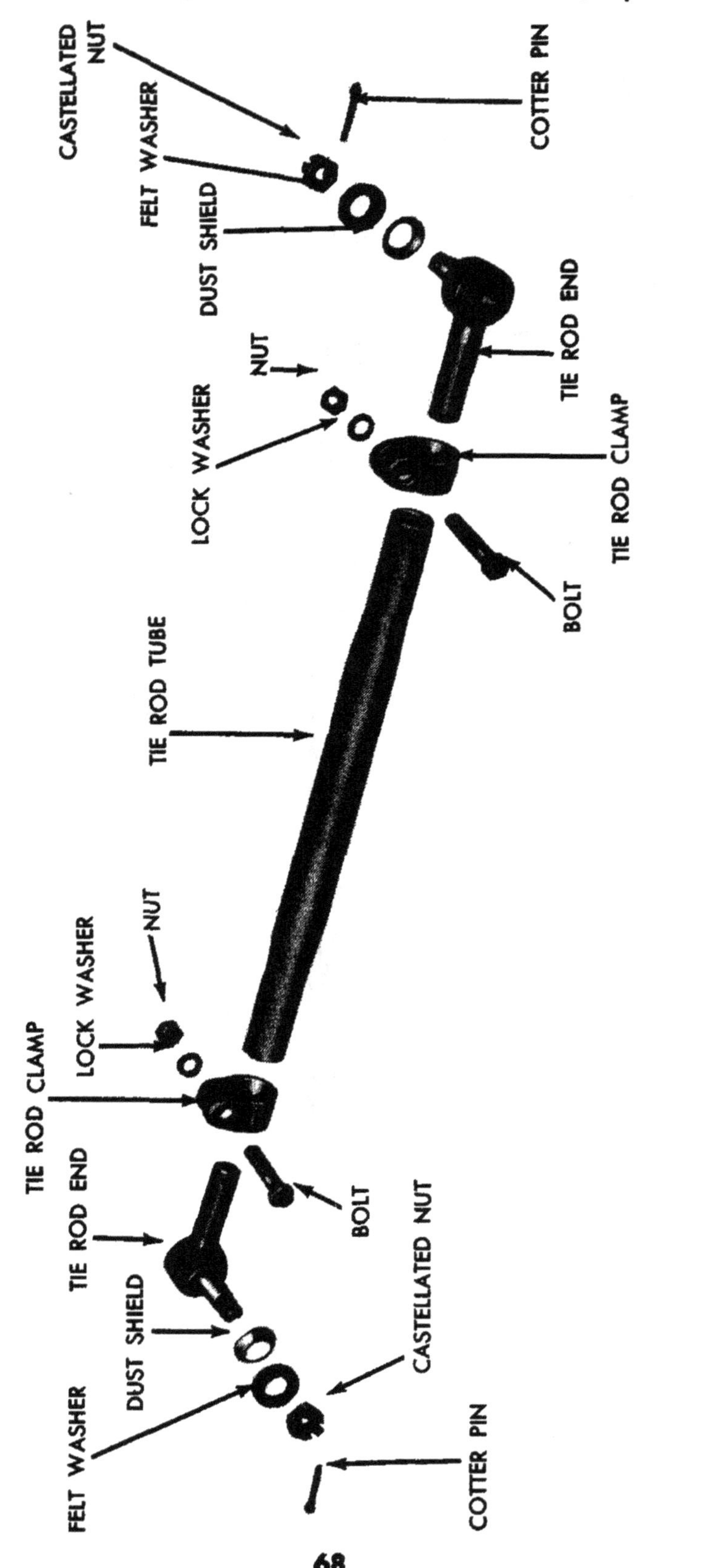

Figure 60 — Tie Rod, Right Side — Exploded View

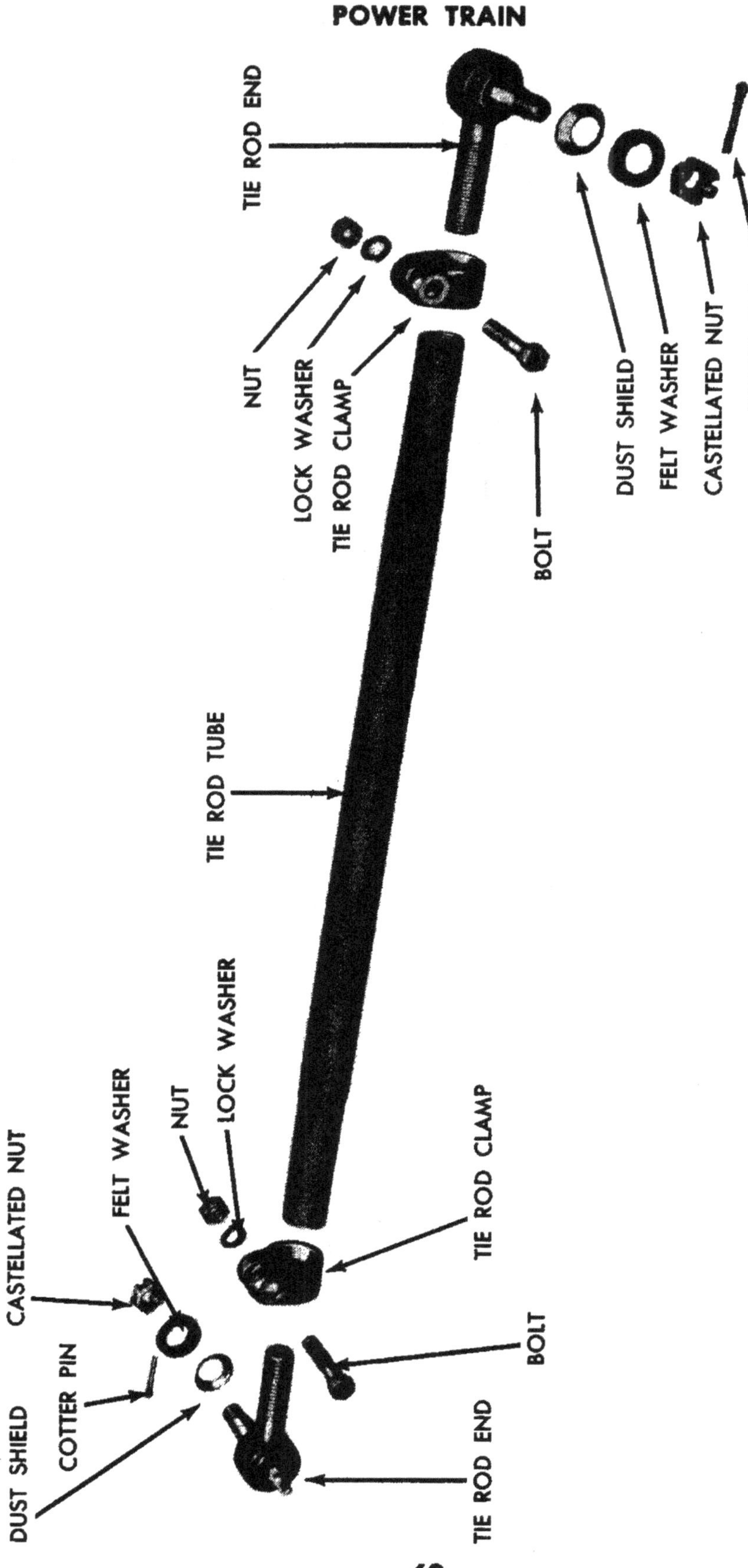

Figure 61 — Tie Rod, Left Side — Exploded View

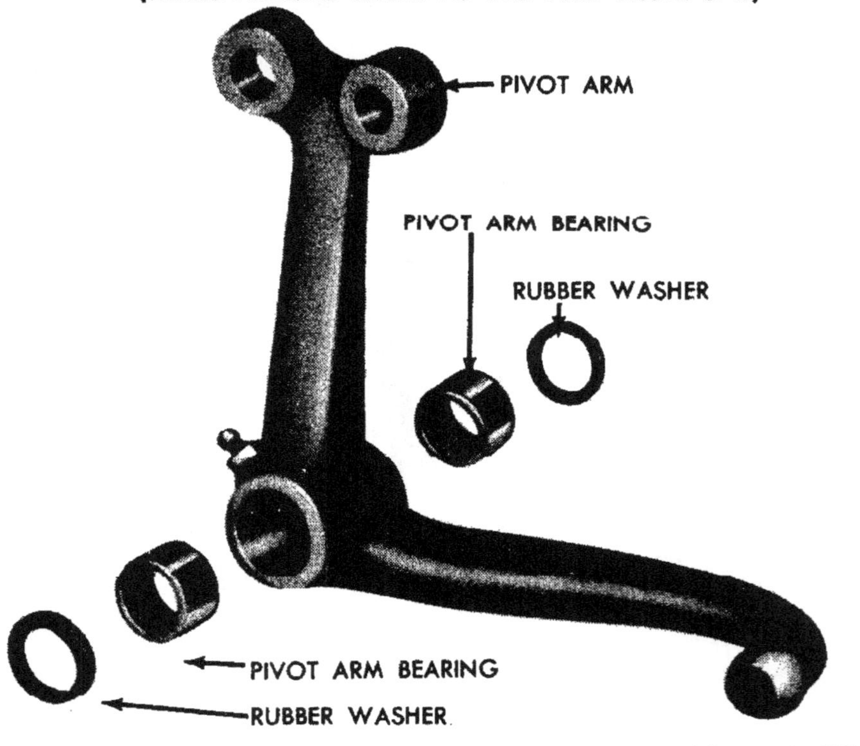

Figure 62 — Pivot Arm — Exploded View

worn ball surfaces. Small nicks or scratches can be removed with a fine stone. Replace universal joint balls, if they are excessively worn or have any flat spots.

(c) Tracta Universal Joint (fig. 45). Replace the inner portion or the outer portion of the axle shafts, if they are bent or have worn splines. Replace the inner portion or the outer portion of the universal joints, if they are cracked or excessively worn. Small nicks or scratches can be removed with a fine stone.

(5) TIE RODS AND PIVOT ARM (figs. 60, 61, and 62).

(a) Inspection. Replace the tie rods if bent or damaged. Replace the tie rod ends if the sockets are loose (step *(b)*, below). Replace the pivot arm, if it is bent or has a worn ball joint. Replace the needle roller bearings in the pivot arm if they are loose or excessively worn (step *(c)*, below).

(b) Tie Rod End Replacement (figs. 60 and 61). Loosen the tie rod clamps at both ends of the tie rod. Remove the tie rod ends from the tie rods. To install tie rod ends, place the tie rod clamps on the tie rod. Install the tie rod ends.

(c) Pivot Arm Needle Bearing Replacement (fig. 62). Place the pivot arm in a press and with a suitable driver, press out the two

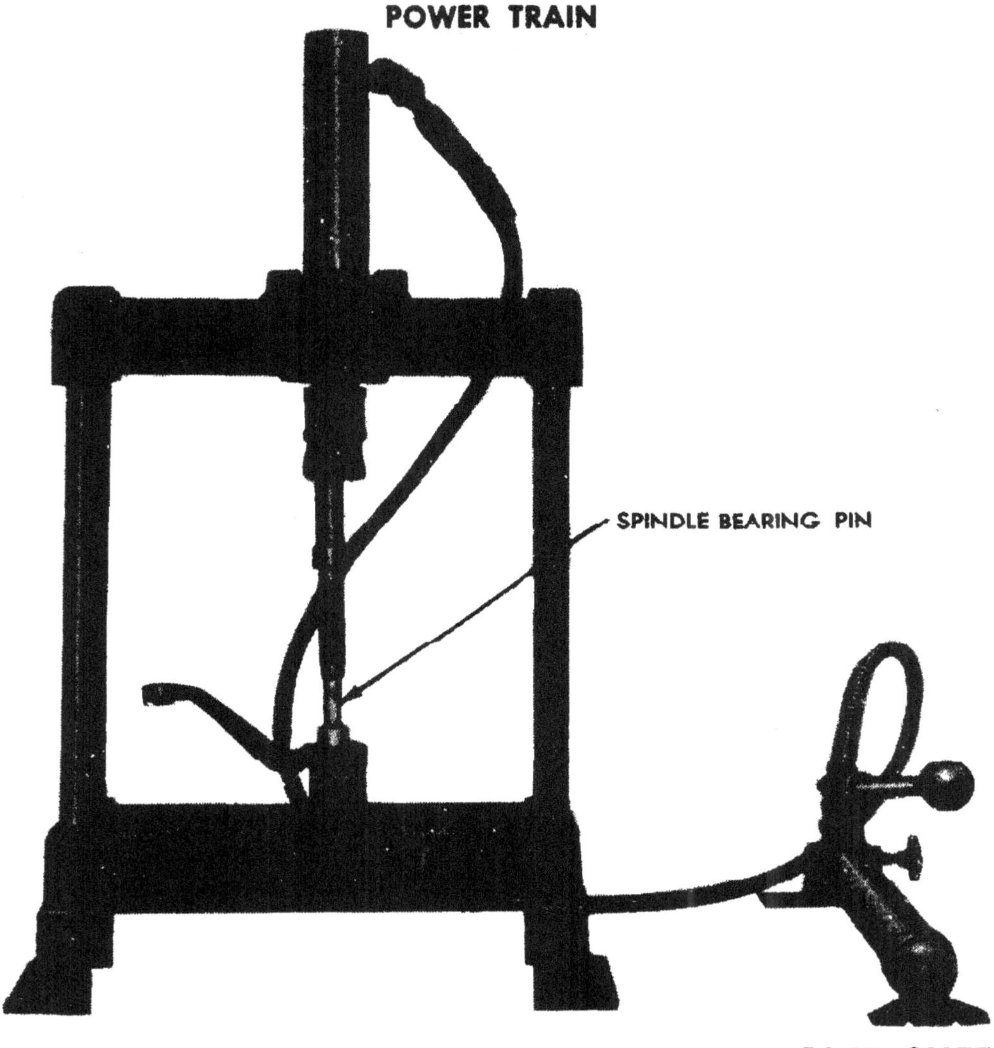

RA PD 28877

Figure 63 — Replacing Spindle Bearing Pin in Spindle Arm

needle bearings. To install the needle bearings, press one needle bearing in the pivot arm about $\frac{1}{16}$ inch below the shoulder of the pivot arm, then turn the pivot arm over and press the other bearing in the arm about $\frac{1}{16}$ inch below the shoulder of the pivot arm.

(6) Spindle Arm and Spindle Housing Bearing Cap.

(a) Inspection. Replace the spindle arm if bent. Replace the spindle arm bearing pin if the diameter of the pin is worn to less than 0.623 inch. Replace the spindle housing bearing cap pin if it is worn to less than 0.625 inch (step *(b)*, below).

(b) Spindle Bearing Pin Replacement. Place the spindle housing bearing cap or the spindle arm (fig. 76) in a press and with a suitable driver, press out the bearing pin. To install a new pin, use a suitable driver and press the pin in until it is flush with the outer

Figure 64 — Removing Spindle Bushing

Figure 65 — Pressing New Bushing in Spindle

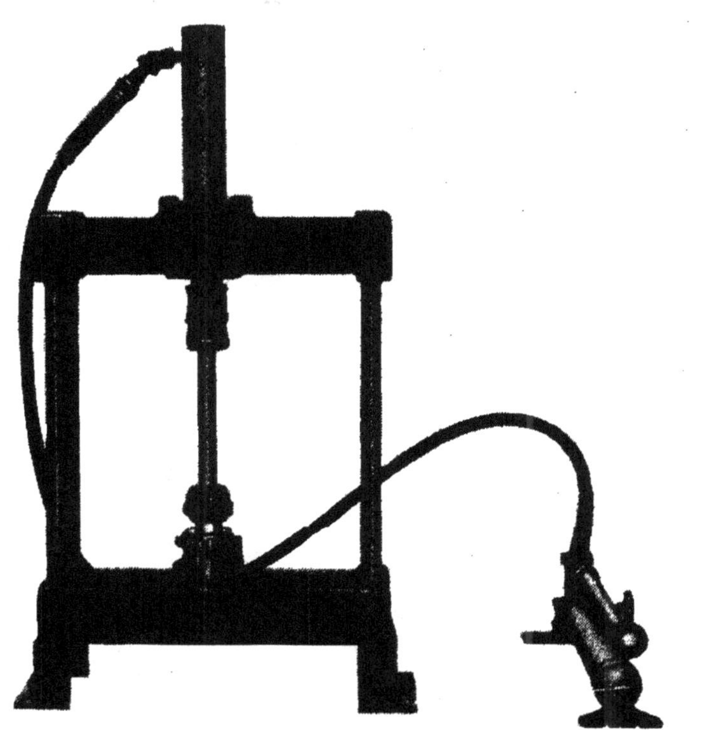

RA PD 28776

Figure 66 — Installing Inner Bearing on Pinion

shoulder. The same procedure applies to both the spindle housing bearing cap and the spindle arm (fig. 63).

(7) SPINDLE HOUSING AND SPINDLE (fig. 76).

(a) Inspection. Replace the spindle housing if it is cracked. If the studs on the spindle housing are bent, broken or damaged, replace them (step *(b)*, below). Replace the spindle, if it has damaged threads or grooved bearing surfaces. If the inner diameter of the spindle bushing is more than 1.225 inch, replace the bushing (subpar. *(c)*, below).

(b) Broken Stud Replacement. Indent the end of the broken stud exactly in the center with a center punch. Drill approximately two-thirds through the broken stud, using a small drill, then follow up with a larger drill (the size of the drill depending on the size of the stud to be removed). The drill selected, however, must leave a wall thicker than the depth of the threads. Select an extractor of the proper size. Insert it into the drilled hole and screw out the remaining part of the broken stud. To install a new stud, use a standard stud driver and drive all studs until no threads show at the bottom of the stud. If the stud is too tight or too loose in the stud hole, select another stud.

(c) Spindle Bushing Replacement. The spindle bushing can be removed with a center punch as shown in figure 64. To install a

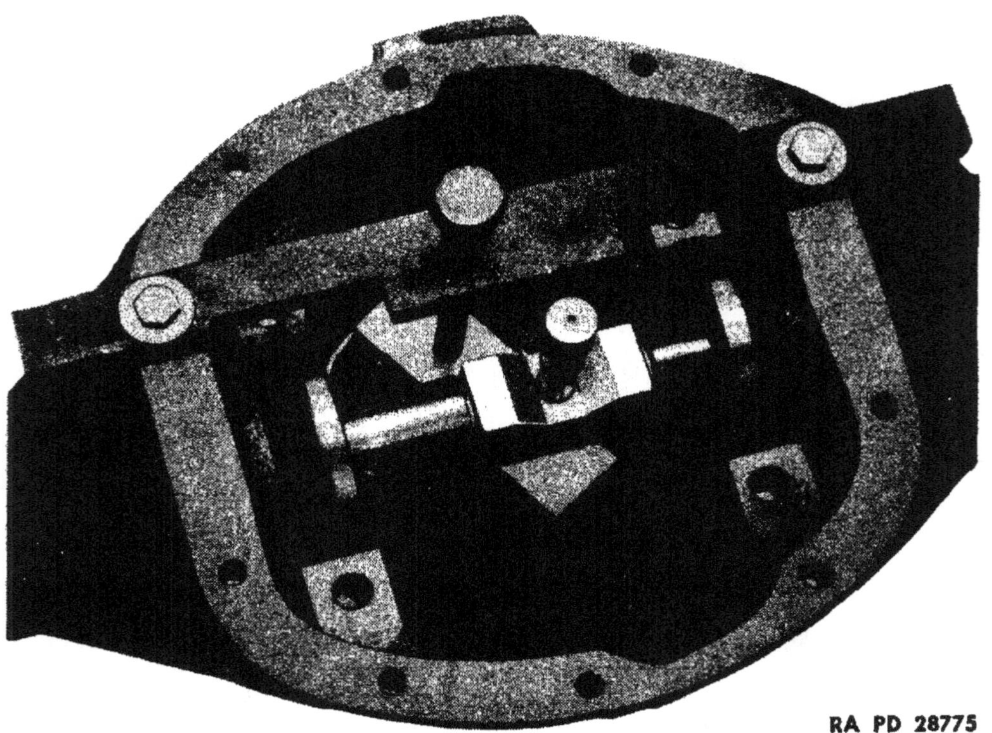

RA PD 28775

Figure 67 — Checking Pinion in Differential Housing With Gage 41-G-176

new bushing, use a suitable driver and press the bushing in the spindle (fig. 65).

26. ASSEMBLY.

a. Install Inner Bearing on Pinion (fig. 66). Press the inner bearing on the pinion, using an arbor press. Make sure the bearing is seated against the shoulder of the pinion gear when installed.

b. Adjust Pinion in Housing (fig. 67). Place the pinion in the differential housing. Install the 41-G-176 gage to check the setting from the back face of the pinion to the center line of the differential case bearing. The standard setting is 0.719 inch. If the gage reading is more than 0.179 inch, shims will have to be added to the inner bearing cup (par. 25 **b** (1) *(c)*). If the gage reading is less than 0.719 inch, shims will have to be removed from the inner bearing cup (par. 25 **b** (1) *(c)*).

c. Install Outer Bearing on Pinion (fig. 57). After the correct pinion setting has been obtained, install the spacer and the original amount of shims on the pinion. If the thickness of the original shims is unknown, install the shims totaling approximately 0.060 inch thick

Figure 68 — Installing Differential Bearing, Using Special Replacer 41-R-2391-65

before installing the outer bearing. Start the outer bearing on the pinion. Install the oil slinger on the pinion.

d. **Adjust the Outer Bearing.** Place the universal joint flange on the pinion. Install the nut on the universal joint flange and draw the flange down tight. Turn the universal joint flange, if there is a slight drag, the pinion bearing adjustment is correct. If the pinion turns with difficulty or can't be turned by hand, shims will have to be added behind the outer bearing. If the pinion turns loosely, shims will have to be removed. If the pinion bearing adjustment is not correct, remove the universal joint flange, and add or remove shims until the correct adjustment is obtained. After the correct adjustment is obtained, again remove the universal joint flange and install the oil seal on the pinion. Install the universal joint flange. Install the nut and cotter pin.

e. **Install Gears in Differential Case** (fig. 58). Place the axle shaft gear thrust washers on the two axle shaft gears. Place the axle shaft gears in the case. Place the two differential pinion thrust washers and gears in the case. Install the differential pinion gear shaft in the case. Install the pinion shaft lock pin in the case and stake the pinion shaft lock pin to prevent the pin from coming out.

Figure 69—Checking Clearance Between Differential Case and Bearing

f. Install Differential Ring Gear (fig. 58). Place the differential ring gear in position on the case. Install the lock plates and cap screws that secure the ring gear to the case. Bend the ears of the lock plates on the cap screws.

g. Install Roller Bearings on Case (fig. 68). If all the original parts have been used in the differential assembly, add the same thickness of shims as originally used and press the roller bearings on the case, then proceed with subparagraph **h**, below. If the original parts are not being used, or if the original shim thickness is not known, install the roller bearings on the case without the shims, and proceed with subparagraph **i**, below.

h. Install Differential Assembly in Housing (fig. 47). Place the bearing cups on the roller bearings. Tilt the bearing cups in order to start the assembly in the housing. Tap the bearing cups lightly until the assembly is seated firmly in the housing. Install the two bearing caps so the numbers on the caps and the housing face the same way and match in every way as shown in figure 47. If the differential assembly being used is not the one originally in the axle, proceed with subparagraph **i**, below.

i. Adjust Differential Assembly (fig. 69). Place the bearing cups on the differential assembly and place the assembly in the housing.

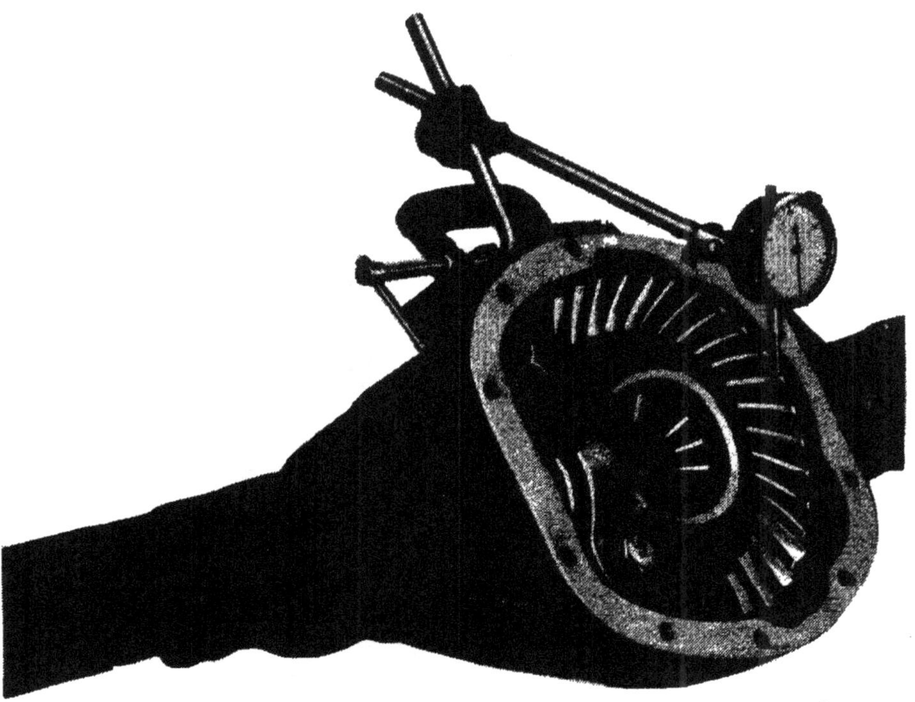

RA PD 28780

Figure 70 — Checking Ring Gear Backlash With Dial Indicator 41-I-100

Slide the assembly to one side of the housing. Check the clearance between the bearing cup and differential housing with a feeler gage. After this clearance has been determined, add 0.008 inch. This will give the thickness of shims required for proper bearing adjustment. Remove the differential assembly from the housing. Remove the bearings from the differential case (par. 24 f (3)). Install the amount of shims determined above in equal amounts on each side of the case and install the bearings back on the case (subpar. g, above). Tilt the bearing cups and place the differential in the housing. Tap the bearing cups lightly until the assembly is seated firmly in the housing. Install the two bearing caps so the numbers on the caps and the housing face the same way, and match in every way.

j. **Check Backlash** (fig. 70). Install a dial indicator 41-I-100 on the differential housing so that the indicator contact is resting on the surface of a ring gear tooth as shown in figure 70. Rotate the ring gear back and forth to determine the backlash. If the backlash is less than 0.005 inch or more than 0.007 inch, remove the differential from the housing (par. 24 e) and remove the bearings from the differential case (par. 24 f (3)). If the backlash was more than 0.007 inch, the ring gear must be brought closer to the pinion. If the backlash was less than 0.005 inch, the ring gear must be moved away from the pinion. This is accomplished by moving shims equal to the

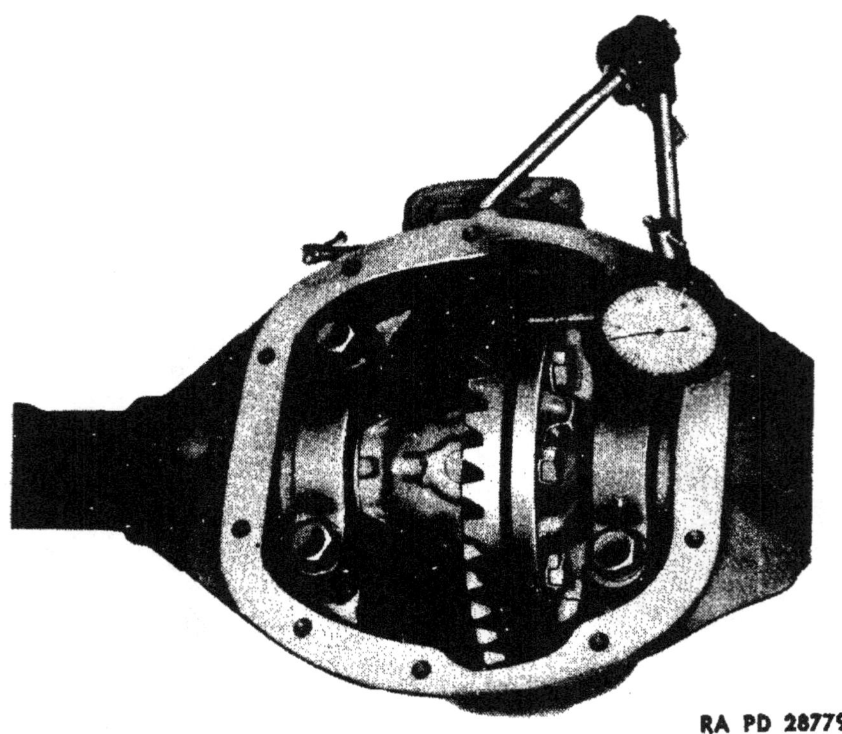

RA PD 28779

Figure 71 —Checking Ring Gear Run-out With Dial Indicator 41-I-100

error in backlash from one side of the case and adding them to the other side. Install the bearings on the case (subpar. g, above). Install differential in housing (subpar. h, above) and recheck the backlash.

k. Check Ring Gear Run-out (fig. 71). Install a dial indicator on the differential housing so that the indicator contact is resting on the flat side of the ring gear as shown in figure 71. Turn the pinion drive flange by hand to determine the run-out on the ring gear. The run-out should not exceed 0.003 inch. If the run-out is more than 0.003 inch, remove the differential assembly from the housing (par. 24 e), and remove the ring gear from the differential case. Check the surface of the differential case and the ring gear for chips or small nicks, which might have occurred during the assembly of the differential. If any small nicks are found, remove them with a fine stone, also check the flange on the differential case for being sprung. Reinstall the differential assembly in the housing (subpar. h, above) and recheck the ring gear run-out.

l. Install Differential Cover (fig. 46). Place a new gasket and the differential cover in place on the axle housing. Install the 10 cap screws that secure the cover to the housing.

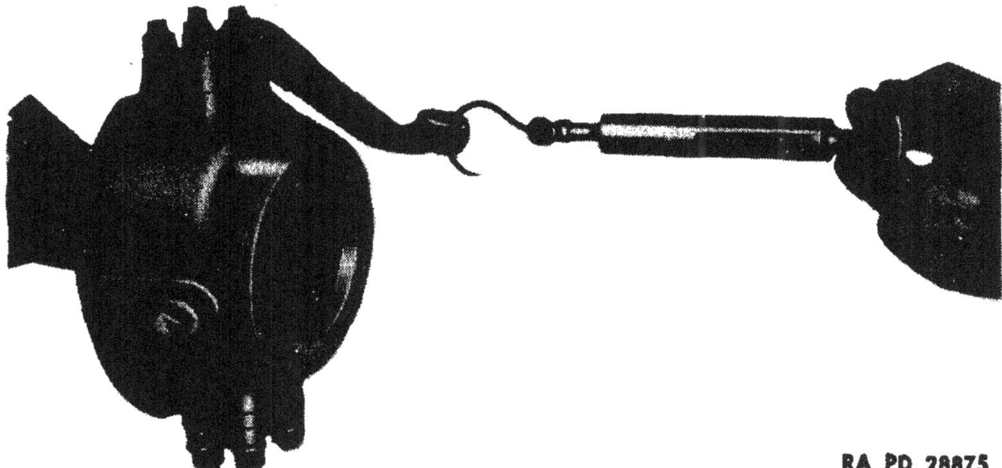

RA PD 28875

Figure 72 — Checking Tension of Spindle Housing

m. Install Pivot Arm (fig. 46). Insert the two rubber seals in the pivot arm. Place the flat washer on the pivot arm shaft. Place the pivot arm on the shaft with the ball joint of the arm facing downward. Place the flat washer and dust shield on the shaft. Install the castellated nut and cotter pin.

n. Install Spindle Housing (fig. 76). Dip the two spindle housing bearings in grease. Place the bearings in the bearing cups on the axle housing. Place the spindle housing on the axle housing with the grease plug to the rear of the vehicle. Install shims totaling 0.048 inch thick on the spindle bearing cap and the spindle arm. Shims are available in thicknesses of 0.003 inch, 0.005 inch, 0.010 inch and 0.030 inch. Install one of each size on the top and bottom of the spindle housing. Place the lower bearing cap on the spindle housing and install the four nuts that secure the cap to the spindle housing. Place the spindle arm on the spindle housing and install the four nuts that secure the spindle arm to the spindle housing.

o. Adjust Spindle Housing (fig. 72). Check the tension of the spindle housing by hooking a scale to the end of the spindle arm. The tension should not be more than 6 pounds or less than 4 pounds. If the tension is over 6 pounds, shims must be removed from the spindle housing. If the tension is less than 4 pounds, shims must be added. When removing or adding shims, be sure the same thickness is removed from, or added to, both ends of the spindle housing. Remove, or add, shims until the correct tension is obtained.

p. Install Spindle Housing Oil Seal (fig. 76). Place a new gasket on the spindle housing. Place the upper and lower halves of the oil seal on the spindle housing. Install the four cap screws that secure the lower half of the oil seal to the spindle housing. Place the axle shaft identification tag on the upper half of the oil seal. Install the four cap screws that secure the upper half of the oil seal to the spindle housing.

q. **Assemble Axle Shafts.** Three different types of front axle universal joints (figs. 38, 42, and 44) are used. Assembly procedures for the Rzeppa and Bendix types are covered in subparagraphs (1) and (2), below. The Tracta type front axle universal joint (fig. 43) requires no assembly before installation (subpar. r (2), below).

(1) RZEPPA JOINT (fig. 38). Hold the cage in a horizontal position and hold the inner race in a vertical position (fig. 41). Insert the inner race in the cage, dropping one of the inner race bosses into one of the larger elongated holes. When the race is entered in the cage, turn the race so that it is entirely in the cage. Line up the two larger elongated holes with the bosses on the axle shaft (fig. 40). Slide the cage in the axle shaft. Holding the cage in this position, insert the thrust washer (fig. 59) behind the cage. Tilt the cage so that it is flush with the shaft. Tilt the cage so that a steel ball can be inserted in the elongated hole (fig. 39). After the steel ball is in position, push the cage down until the opposite side of the cage is exposed. Insert another steel ball in the elongated hole on the cage and push the cage down. Repeat this operation until all the steel balls are in the cage. Insert the pilot pin (fig. 59) in position. Insert the retainer on the axle shaft and secure the retainer with the snap ring. Insert the inner shaft in the outer shaft. Install the three flat head screws that secure the retainer to the inner race.

(2) BENDIX JOINT (figs. 42 and 43). Place the universal joint knuckle in an upright position in a vise. Insert the center ball in the hole of the knuckle. Place the center ball in its race on the center ball pin hole. Arrange the center ball so that the grooved side of the center ball is away from the pin hole. Insert the three universal joint balls in their races. Arrange the center ball so that the grooved side is in line with the race of the last ball to be installed as shown in figure 42 and drop the ball in its race. Rotate the center ball in its race until the hole in the ball is in line with the center ball pin. Remove the assembly from the vise. Turn the assembly over so that the pin may drop in the hole of the center ball. Install the grooved pin in the knuckle and stake the pin to prevent it from coming out.

r. **Install Axle Shaft.**

(1) BENDIX AND RZEPPA JOINTS. Slide the axle shaft in the axle housing. It will be necessary to turn the axle shaft until the splines on the axle shaft are in line with the axle shaft gear in the differential.

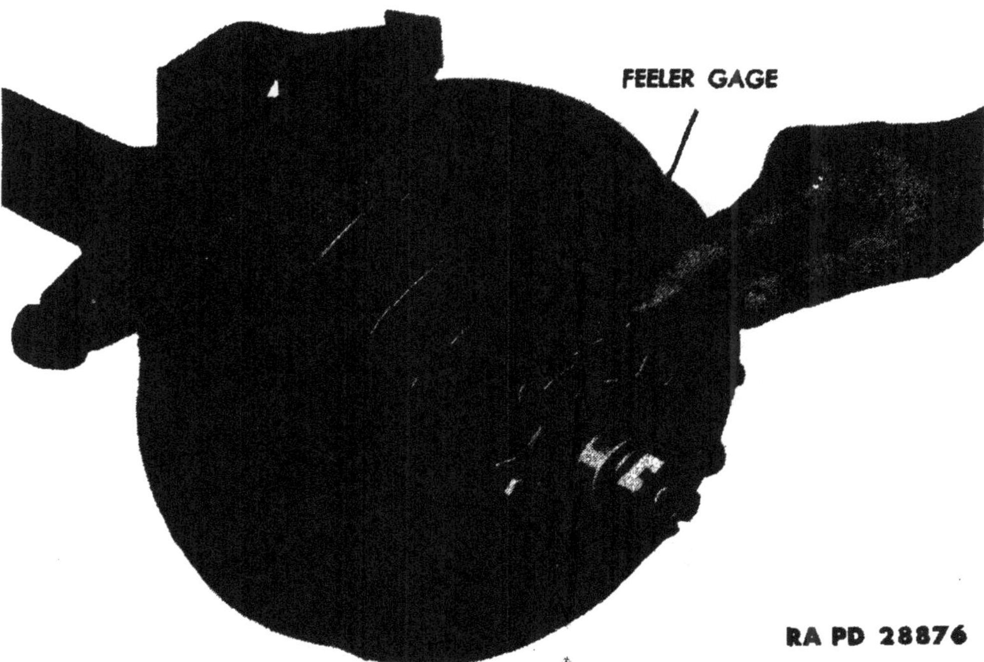

Figure 73 — Checking Clearance Between Drive Flange and Hub

(2) TRACTA JOINT (fig. 45). Slide the inner portion of the axle shaft and the inner portion of the universal joint into the axle housing. Turn the axle shaft so as to line up the splines of the axle shaft with the axle shaft gear in the differential. Slide the outer portion of the universal joint on the outer portion of the axle shaft. Line up the slots of the .two universal joints and slide the outer axle shaft in place on the axle.

s. **Install Brake Plate and Spindle.** Place the spindle on the spindle housing. Place the brake plate on the spindle with the wheel cylinder toward the top of the brake plate. Line up the holes in the brake plate and spindle with the spindle housing. Install the six cap screws that secure them to the spindle housing.

t. **Install Hydraulic Brake Hose** (fig. 46). Install the brake hose to the brake line on the axle housing. Install the clamp to the brake hose at the bracket on the axle housing. Insert the brake hose through the guard and connect the hose to the brake line on the brake plate. Install the brake hose clamp at the guard.

u. **Install Hub and Brake Drum** (fig. 46). Pack the wheel bearings with the specified lubricant. Insert the hub and brake drum on the spindle with the inner wheel bearing and grease retainer in the hub. Insert the smaller thrust washer on the spindle and install the bearing adjusting nut. Tighten the adjusting nut until the brake

855473 O - 49 - 6

RA PD 28868

Figure 74 — Checking Wheels With Straightedge

drum binds when turned; then back off the adjusting nut one-eighth turn. This will give the correct wheel bearing adjustment. Install the lock washer and lock nut on the spindle. Bend the ears of the lock washer over the lock nut.

v. **Install Drive Flange** (fig. 73).

(1) RZEPPA TYPE AXLE SHAFTS. Install a 0.060-inch thickness of shims between the drive flange and the hub. Place the drive flange on the axle shaft. Install the six cap screws that secure the drive flange to the hub. Install the castellated nut on the axle shaft. Install the hub cap on the drive flange.

(2) BENDIX OR TRACTA TYPE AXLE SHAFT (fig. 73). Place the drive flange on the axle shaft. Install the castellated nut on the axle shaft and draw it down tight. Turn the front wheels to the maximum left or right and measure the space between the drive flange and hub with a feeler gage (fig. 73) to determine the number of shims to be installed. Remove the nut from the axle shaft and remove the drive flange. Add the required thickness of shims between the drive flange and the hub. Install the six cap screws that secure the drive flange to the hub. Install the castellated nut on the axle shaft. Back off the nut on the axle shaft until a 0.50-inch feeler gage can pass between the nut and drive flange. Tap the nut on the axle shaft lightly. The axle shaft will move inward. Again check the space between the nut and drive flange. The space should not be less than 0.015 inch or more than 0.035 inch. If the space is less than 0.015 inch, add shims behind the drive flange and hub until this limit is obtained. If the space is more than 0.035 inch, remove shims from the drive flange until the above limit is obtained. Draw the nut on the axle shaft up tight. Install the hub cap.

w **Install Tie Rods** (fig. 46). Insert the ends of the tie rods in the spindle arms and pivot arm. Be sure the dust shield and felt washer are on the tie rod ends. Install the castellated nuts that secure the tie rod ends to the spindle arms and to the pivot arm.

27. INSTALLATION.

a. **Preliminary Work.** Place a hydraulic jack under the front axle assembly. Roll the assembly under the vehicle. Raise the assembly until the front springs can be raised and secure to the spring shackles. Lower the jack to allow the axle assembly to rest on the springs.

b. **Install Spring U-bolts** (fig. 35). Place the spring U-bolts in position on the axle housing. Install the spring seat plate on the U-bolts at the right side of the axle. Install the four nuts that secure the spring seat to the U-bolts. Raise the torque reaction spring in position on the U-bolts on the left side. Install the nuts that secure the torque reaction spring to the U-bolts.

RA PD 28869

Figure 75 — Adjusting Toe-in, Using Wheel Alining Gage 41-G-510

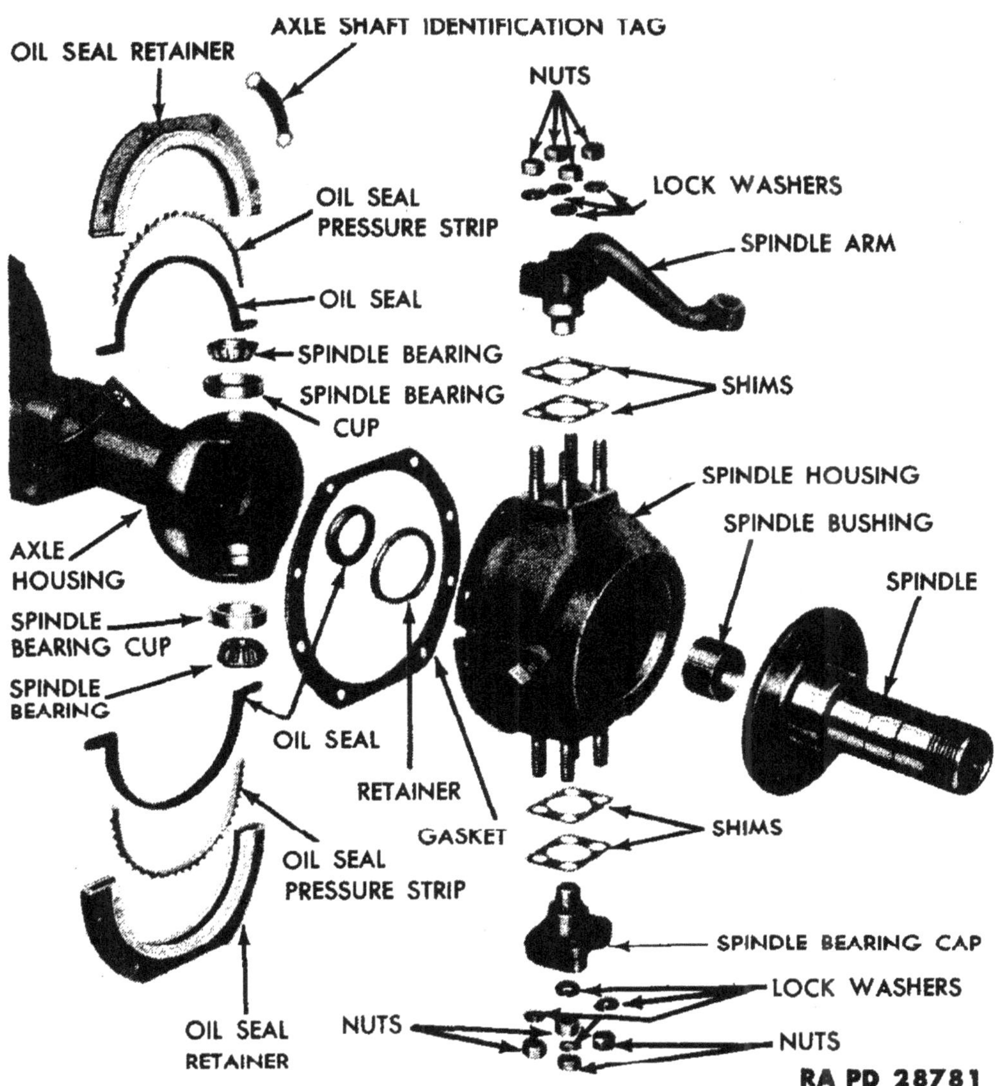

Figure 76 — Spindle Housing — Exploded View

c. **Install Shock Absorbers** (fig. 35). Insert a rubber mounting in each side of each shock absorber eye. Place the shock absorber on the mounting bracket at the spring seat plate. If new rubber mountings are being used, compress them with compressor 41-C-2554-400. Install the flat washer and cotter pin that secure the shock absorber to the spring seat plate. Place the left shock absorber on the mounting bracket at the torque reaction spring. Install the flat washer and cotter pin that secure the shock absorber to the torque reaction spring.

d. **Install Propeller Shaft, Drag Link, and Wheels** (fig. 35). Install the propeller shaft to the front axle (par. 21 **b**). Place the drag link in the ball joint on the pivot arm. Install the drag link plug on the drag link. Install the cotter pin in the drag link plug. Install the wheels.

e. **Lubricate.** Fill the differential to proper level with specified oil. Apply specified grease in each spindle housing and to all fittings. Bleed the hydraulic brake system. Refer to TM 9-803.

28. WHEEL ALINEMENT.

a. **Caster and Camber.** The caster and camber is established at the time of manufacture and cannot be altered by any adjustment.

b. **Toe-in.**

(1) EQUALIZE TIE RODS (fig. 74). Set the pivot arm parallel to the axle and place a straightedge against the left rear and left front wheel. If the rear or front sides of the front tire do not touch the straightedge, the tie rods must be adjusted. Loosen the tie rod clamps at both ends of the left tie rod. Turn the tie rod tube clockwise to bring the forward side of the front wheel inward, or counterclockwise to bring the rear side of the front wheel inward. Adjust until the straightedge touches the side of the front tire at both front and rear. Repeat this procedure on the right-hand side of the vehicle.

(2) ADJUST TOE-IN (fig. 75). After the tie rods have been equalized, pull the vehicle forward at least three feet to remove the backlash and place the telescoping wheel alining gage 41-G-510 between the wheels in front of the axle so that the chains on both ends of the gage are barely touching the floor. Set the scale so the pointer registers zero. Pull the vehicle forward until the gage is brought to a position back of the axle with both chains barely touching the floor. The reading at this point will be the amount of toe-in or toe-out. Adjust the right-hand tie rod until a toe-in of 1/16 inch is obtained. Recheck the toe-in after making the adjustment. Tighten the tie rod clamps.

Section VI

REAR AXLE

29. DESCRIPTION AND DATA.

a. **Description** (fig. 78). The rear axle is the full-floating type, designed so that the axle shafts can be removed without disturbing the wheels. The differential drive is of the hypoid type. The differential parts are identical and are interchangeable with the front axle.

b. Data.

Rear Axle:

Type	Full-floating
Make	Spicer
Drive	Through springs
Road clearance	8⅞ in.

Differential:

Type	Hypoid
Ratio	4.88 to 1
Bearings	Timken roller
Oil capacity (pt)	2.5

Pinion Shaft:

Bearings	Timken
Adjustment	Shims
Backlash	0.005 to 0.007 in.

30. REMOVAL.

a. Preliminary Work. Remove the differential drain plug and drain the oil. Raise the rear of the vehicle until the weight of the vehicle is off the rear springs.

b. Disconnect Propeller Shaft (fig. 77). Remove the four nuts and two U-bolts that secure the propeller shaft to the universal joint flange at the rear axle. Slide the propeller shaft off the universal joint flange. Wrap a piece of tape around the bearings on the propeller shaft to prevent losing the bearings.

c. Disconnect Shock Absorbers and Hydraulic Brake Line (fig. 77). Remove the cotter pin and flat washer that secure the two rear shock absorbers to the bracket on the spring seat plates. Pull the shock absorbers off the bracket. Disconnect the hydraulic brake line leading to the rear axle at the differential housing.

d. Remove Spring U-bolts (fig. 77). Remove the four nuts that secure the spring U-bolts at both rear springs. Remove the U-bolts and the spring seat plates from the axle.

e. Disconnect Springs (fig. 77). Remove the lower spring shackle bushings at the rear of both springs. Pull both springs off the spring shackles. Drop the springs to the floor and roll the rear axle assembly out from the vehicle.

RA PD 329206

Figure 77 — Rear Axle Assembly in Vehicle

31. DISASSEMBLY.

a. **Remove Wheels.** Remove the five nuts that secure each wheel to the hub. Remove the wheels.

b. **Remove Axle Shafts** (fig. 82). Remove the six cap screws that secure the drive flange to the hub. Install two of the cap screws that were removed from the drive flange in the two threaded holes on the drive flange. Draw the cap screws down until the drive flange is free from the hub. Remove the axle shafts from the axle housing.

c. **Remove Hub and Drum Assembly** (fig. 82). Bend the ears of the flat washer off the bearing lock nut. Remove the bearing lock nut and bearing adjusting nut off the housing, using the wrench furnished with the vehicle. Slide the hub and drum assembly with the wheel bearings off the housing.

d. **Remove Brake Plate** (fig. 82). Disconnect the hydraulic brake line at the brake plate. Remove the six cap screws that secure the brake plate to the axle housing. Remove the brake plate from the axle housing.

e. **Remove Differential Assembly.** Remove the 10 cap screws that secure the differential cover to the housing (fig. 78). Remove the differential cover. Remove the 4 cap screws from the 2 differential bearing caps (fig. 47), and remove the caps. Remove the differential assembly from the housing, using a pry bar if necessary. Reinstall the bearing caps in the housing, noting the markings (fig. 47) to assure their being installed in their correct location.

f. **Remove Differential Pinion Gears and Axle Shaft Gears From Differential** (fig. 48). Place the differential assembly in a vise equipped with brass jaws. With a long-nosed drift, drive the differential pinion shaft tapered pin out of the differential gear case (fig. 48). Tap the differential pinion shaft from the case with a brass drift and hammer. Remove the two differential pinion gears and thrust washers and the two axle shaft gears and thrust washers from the case.

g. **Remove Ring Gear From Case** (fig. 58). Bend the ears of the lock plates off the cap screws. Remove the cap screws that secure the ring gear to the case, and remove the ring gear.

h. **Remove Roller Bearings From Differential Case** (fig. 49). Place the differential case in a vise. Install the bearing remover 41-R-2378-30 to the roller bearing. Remove the roller bearings from both ends of the differential case. Remove the shims. Note the thickness of the shims removed from each side to assist in reassembly.

i. **Remove Drive Pinion.** Remove the nut and flat washer that secure the universal joint axle end flange to the drive pinion. Install

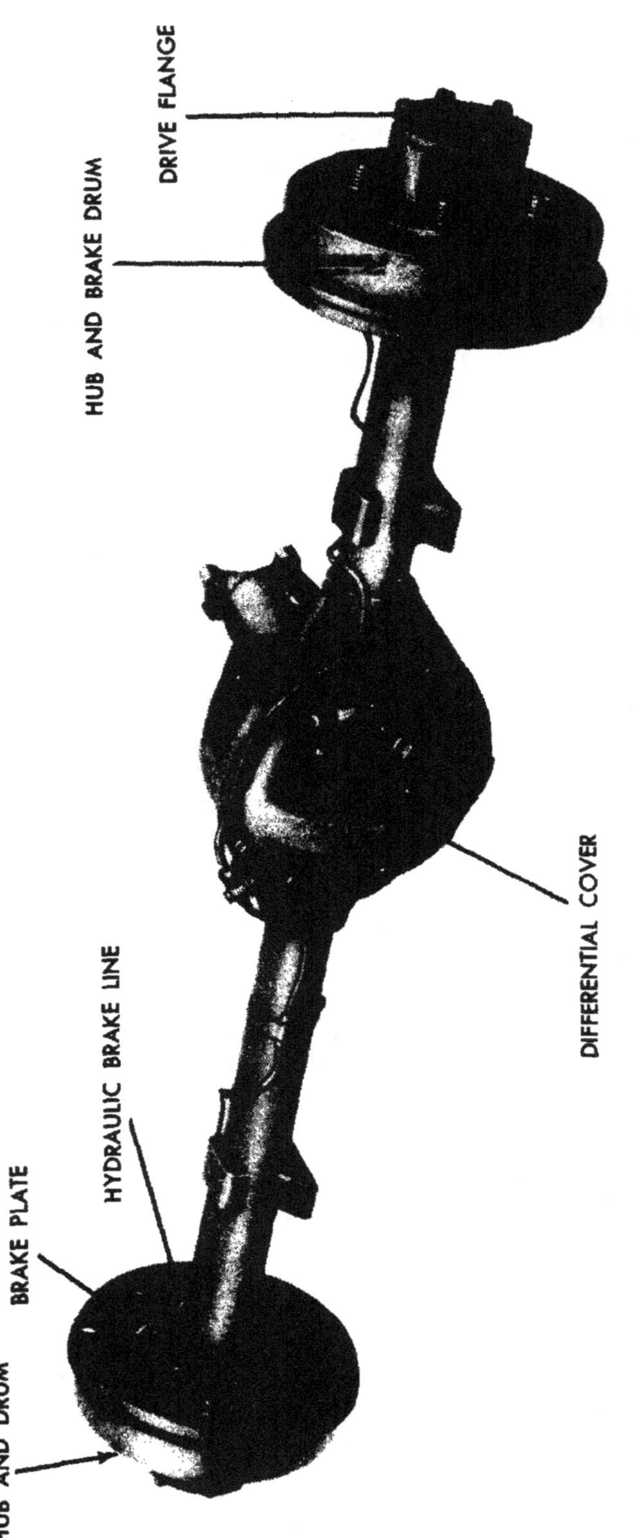

Figure 78 — Rear Axle Assembly

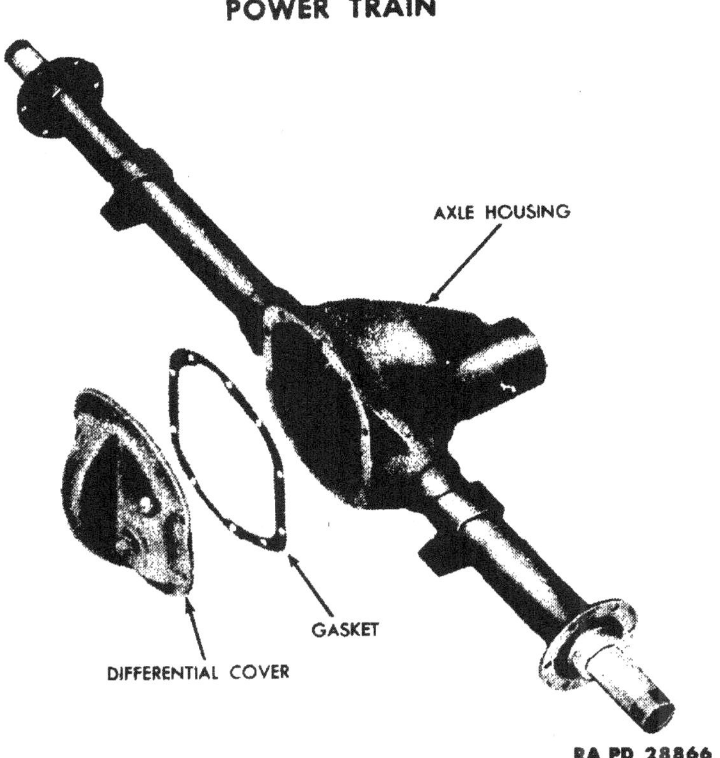

Figure 79 — Rear Axle Housing

the flange puller to the universal joint flange (fig. 50) and remove the flange. Using a brass drift and hammer, drive the drive pinion out of the axle housing (fig. 51). Remove the shims and spacer from the drive pinion. Note the thickness of shims removed from the pinion to assist in reassembly.

32. CLEANING, INSPECTION AND REPAIR.

a. Cleaning. Clean all parts in dry-cleaning solvent. Rotate the bearings in dry-cleaning solvent until all trace of lubricant has been removed. Oil the bearings immediately to prevent corrosion of the highly polished surface.

b. Inspection and Repair.

(1) AXLE HOUSING AND COVER (fig. 79).

(a) Inspection. Replace the axle housing if it is broken at any of the welds or if it is cracked or bent. Replace the drive pinion bearing cups if they are pitted, corroded, or discolored due to overheating (subpar. *(b)*, below). Replace the oil seals in the axle housing regardless of their condition (step *(c)*, below). Replace the differential cover if cracked or if it has damaged threads in the filler plug hole. Replace the breather cap on the cover, if it is missing or damaged.

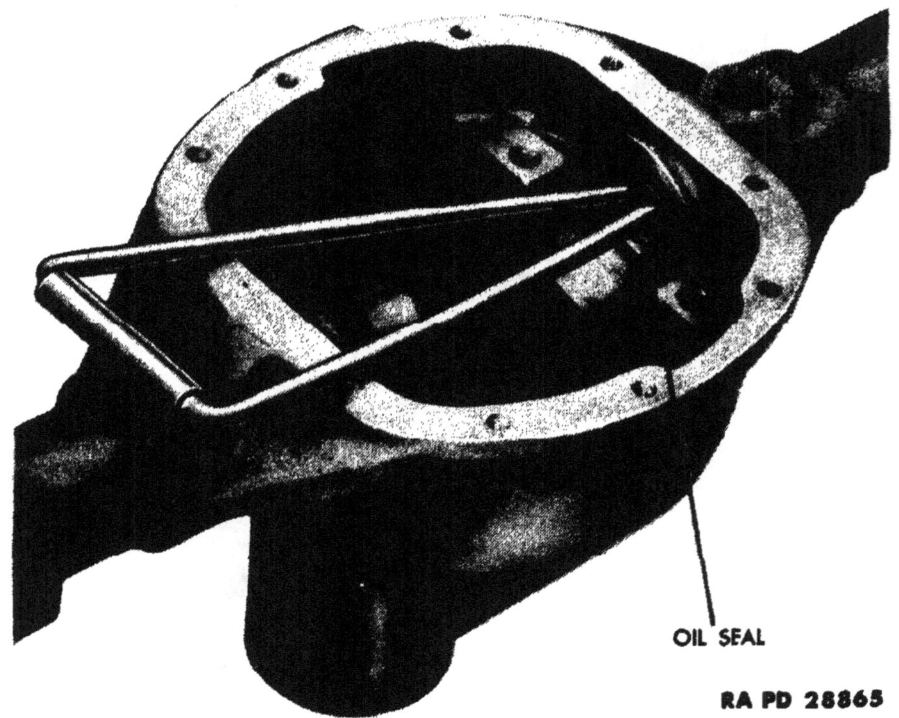

Figure 80 — Removing Oil Seal From Axle Housing, With Remover 41-R-2384-38

(*b*) *Drive Pinion Bearing Cap Replacement.* Remove the inner and outer bearing cups, using a standard puller. To assist in assembly, note the thickness of shims when removing the inner bearing cup. To install a new bearing cup, use a brass drift and hammer. Place the original thickness of shims behind the inner bearing cup and tap the bearing cup lightly around the entire circumference of the cup until it is flush with the shoulder in the axle housing (fig. 52).

(*c*) *Oil Seal Replacement* (fig. 80). Remove the inner oil seal with the remover 41-R-2384-38. To install a new oil seal, use special replacer 41-R-2391-20 and tap the oil seals in place (fig. 81).

(2) DRIVE PINION ASSEMBLY (fig. 57). Replace any roller bearings that are pitted, corroded, or discolored due to overheating. Replace the drive pinion gear if it has excessively worn, or broken teeth, or if the splines are worn or the threads damaged. The differential gear and the drive pinion are furnished in matched sets only, and if either is found damaged, both must be replaced. Small nicks can be removed from the pinion gear with a fine stone.

(3) DIFFERENTIAL ASSEMBLY (fig. 58). Replace any gears that are excessively worn or have any missing teeth. The differential ring gear and the drive pinion are furnished in matched sets only, and if

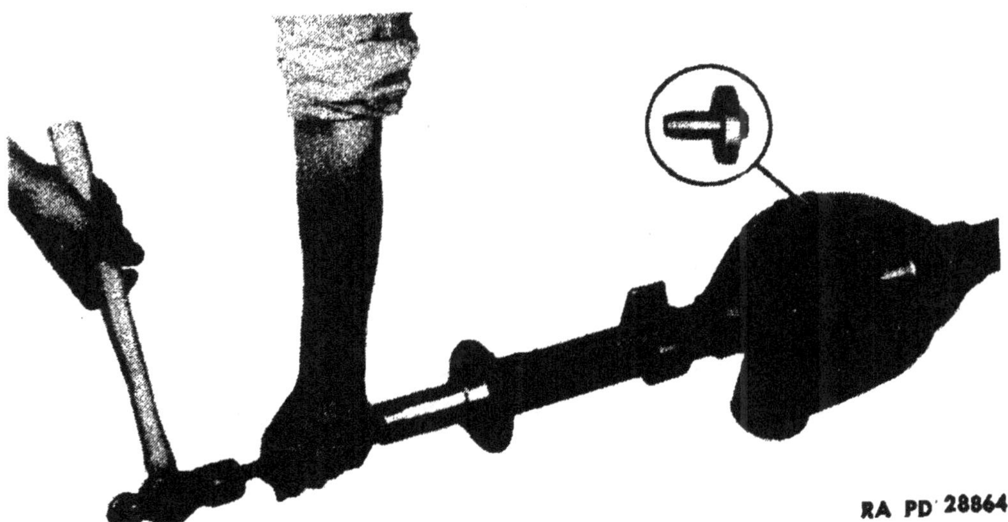

RA PD 28864

Figure 81 — Installing Oil Seal With Replacer 41-R-2391-20

either is found damaged, both must be replaced. Replace the differential pinion gear if its inside diameter is more than 0.625 inch. Replace the differential pinion shaft, if the inside diameter is worn to less than 0.625 inch. Replace the axle shaft gear if the hub is worn to less than 1.500 inches. Replace the differential pinion gear and the axle shaft gear thrust washer if the thickness is worn to less than 0.032 inch. Roller bearings and cups that are pitted, corroded, or discolored due to overheating must be replaced.

(4) AXLE SHAFT (fig. 82). Replace the axle shafts if they are bent or have any worn or broken splines.

33. ASSEMBLY.

a. **Install Inner Bearing on Pinion** (fig. 66). Press the inner bearing on the pinion, using an arbor press. Make sure the bearing is firmly seated on the shoulder of the pinion gear when installed.

b. **Adjust Pinion in Housing** (fig. 67). Place the pinion in the differential housing. Install the gage 41-G-176 to check the setting from the back face of the pinion to the center line of the differential case bearing. The standard setting is 0.719 inch. If the gage reading is more than 0.719 inch, shims will have to be added to the inner bearing cup (par. 32 **b**). If the reading is less than 0.719 inch, shims will have to be removed from the inner bearing cup (par. 32 **b**).

c. **Install Outer Bearing on Pinion** (fig. 57). After the correct pinion setting has been obtained, install the spacer and the original amount of shims on the pinion. If the thickness of the original shims is unknown, install shims totaling approximately 0.060 inch thick.

Start the outer bearing on the pinion. Install the oil slinger on the pinion.

d. Adjust the Outer Bearing. Place the universal joint flange on the pinion. Install the nut on the pinion and draw the universal joint flange down tight. Turn the universal joint flange. If there is a slight drag, the pinion bearing adjustment is correct. If the pinion turns with difficulty or cannot be turned by hand, shims should be added behind the outer bearing. If the pinion is too loose, shims should be removed. Remove the universal joint flange and add, or remove, shims, until the correct adjustment is obtained. After the correct adjustment is obtained, again remove the universal joint flange and install the oil seal on the pinion. Install the universal joint flange. Install the nut and cotter pin.

e. Install Gears in Differential Case (fig. 58). Place the axle shaft gear thrust washers on the two axle shaft gears. Place the axle shaft gears in the case. Place the two differential pinion thrust washers and gears in the case. Install the differential pinion gear shaft that secures the two differential pinion gears in the case. Install the pinion shaft lock pin in the case.

f. Install Differential Ring Gear (fig. 58). Place the differential ring gear in position on the case. Install the lock plates and cap screws that secure the ring gear to the case. Bend the ears of the lock plate on the cap screws.

g. Install Roller Bearings on Case (fig. 68). If all the original parts have been used in the differential assembly, add the same thickness of shims as originally used, and press the roller bearings on the case, then proceed with subparagraph **h**, below. If the original parts are not being used, or if the original shim thickness is not known, install the roller bearings on the case without the shims, and proceed with subparagraph **i**, below.

h. Install Differential Assembly in Housing (fig. 47). Place the bearing cups on the bearings. Tilt the bearing cups in order to start the assembly in the housing. Tap the bearing cups lightly until the assembly is seated firmly in the housing. Install the two bearing caps so that the numbers on the caps, and the housing face the same way, and match in every way as shown in figure 47. If the differential assembly being used is not the one originally in the axle, proceed with subparagraph **i**, below.

i. Differential Assembly Adjustment (fig. 69). Place the differential assembly in the housing with the bearing cups on the assembly. Slide the assembly to one side of the housing. Check the clearance between the bearing cup and differential housing with a feeler gage. After this clearance has been determined, add 0.008 inch. This will give the thickness of shims required for proper bearing adjustment. Remove the differential assembly from the housing. Remove the bearings from the differential case (par. 24 f (3)). Install the number of shims, determined above, on each side of the case and install the bearings back on the case (par. 26 e (3)). Tilt the bearing cups in order to start the assembly in the housing. Tap the bearing cups lightly until the assembly is seated firmly in the housing. Install the two bearing caps so the numbers on the bearing caps and housing face the same way and match in every way.

j. **Check Backlash** (fig. 70). Install a dial indicator on the differential housing so that the indicator contact is resting on the surface of a ring gear tooth as shown in figure 70. Rotate the ring gear back and forth to determine the backlash. If the backlash is less than 0.005 inch or more than 0.007 inch, remove the differential from the housing (par. 24 e) and remove the bearings from the differential case (par. 24 f (3)). If the backlash was more than 0.007 inch, the ring gear must be brought closer to the pinion. If the backlash was less than 0.005 inch, the ring gear must be moved away from the pinion. This is accomplished by removing the shims, equal to the error in backlash, from one side of the case, and adding them to the other side of the case. Install the bearings on the case (subpar. g, above). Install the differential in the housing (subpar. h, above).

k. **Check Ring Gear Run-out** (fig. 71). Install a dial indicator on the differential housing so that the indicator contact is resting on the flat side of the ring gear as shown in figure 71. Turn the pinion drive flange by hand to determine the run-out of the ring gear. The run-out should not exceed 0.003 inch. If the run-out is more than 0.003 inch, remove the differential assembly from the housing (par. 24 e) and remove the ring gear from the differential case. Check the surface of the differential case and the ring gear for chips or small nicks which might have occurred during the assembly of the differential. If any small nicks are found, remove them with a fine stone; also check the flange on the differential case for being sprung. Reinstall the differential assembly in the housing (subpar. h, above) and recheck the ring gear run-out.

l. **Install Differential Cover** (fig. 78). Place a new gasket and the differential cover in place on the axle housing. Install the ten cap screws that secure the cover to the housing.

m. **Install Brake Plate** (fig. 82). Place the brake plate on the housing, with the brake cylinder on the brake plate toward the top. Line up the holes in the brake plate with the axle housing. Install the six cap screws that secure it to the axle housing. Install the hydraulic

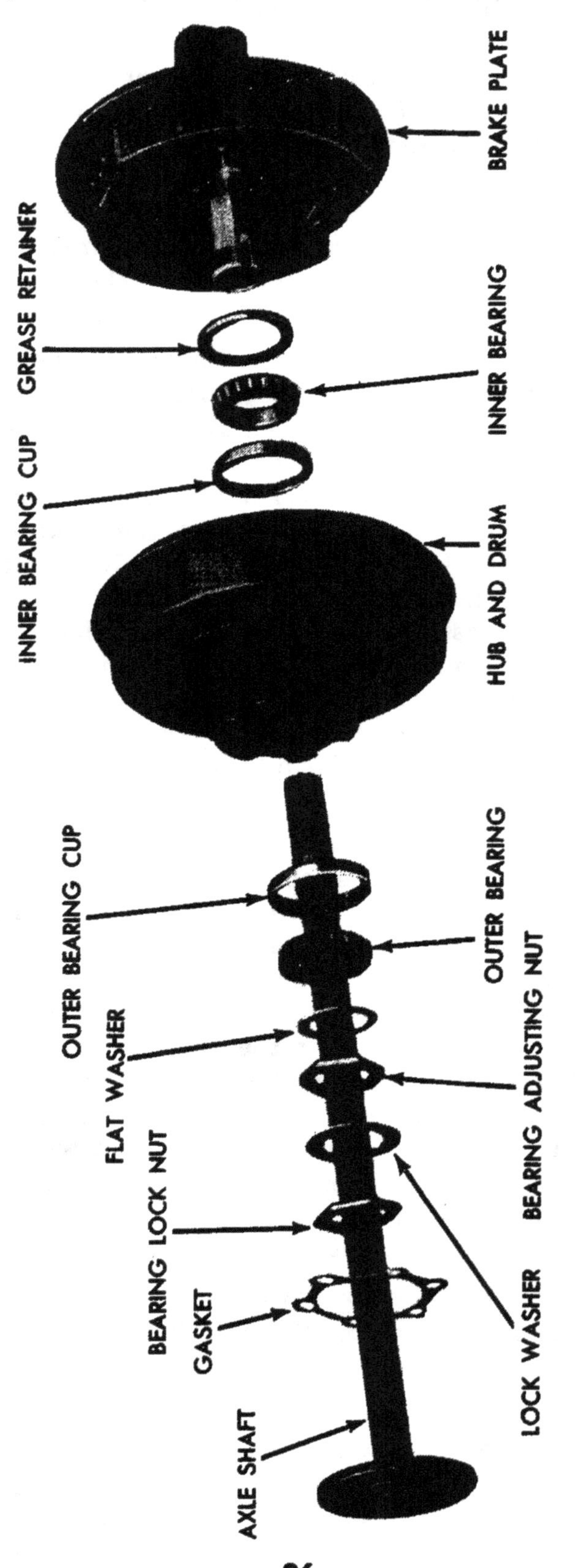

Figure 82 — Axle Shaft — Exploded View

brake line at the connection on the brake plate. Install the flexible hydraulic brake line leading from the frame crossmember at the connection at the differential housing.

n. **Install Hub and Brake Drum** (fig. 82). Pack the wheel bearings with the specified lubricant. Install the inner bearing in place in the hub and install the hub and brake drum on the housing. Install the outer wheel bearing and thrust washers. Install and tighten the bearing adjusting nut until the brake drum binds, then back it off one-sixteenth turn. This will give the correct wheel bearing adjustment. Install the lock washer and lock nut. Bend the ears of the lock washer over the lock nut.

o. **Install Axle Shafts** (fig. 82). Insert the axle shaft in the axle housing. Turn the axle shaft to line up the splines on the axle shaft with the gear in the differential. Install the six cap screws that secure the drive flange to the hub.

p. **Install Wheels.** Place the wheel in position on the hub and secure it with five cap screws.

34. INSTALLATION.

a. **Preliminary Work.** Place the rear axle assembly under the vehicle. With a hydraulic jack, raise the rear axle high enough so that the spring shackles can be connected.

b. **Install Springs** (fig. 77). Raise the two rear springs and install them on the spring shackles. Install the spring shackle bushings in the spring shackles. Lower the jack until the axle assembly is resting on the springs, making sure that the spring tie bolt is in line with the hole on the axle housing.

c. **Install Spring U-bolts** (fig. 77). Place the spring U-bolts in position on the axle housing. Install the spring seat plate on the U-bolts and secure it to the spring with four nuts. The same procedure applies for installing the U-bolts on the other spring.

d. **Install Shock Absorbers** (fig. 77). Insert a rubber mounting in each side of each shock absorber eye. Place the lower end of the shock absorber on the bracket at the spring seat plate. If new shock absorber rubber mountings are being used, compress them with compressor 41-C-2554-400. Install the flat washer and cotter pin that secure the shock absorber to the bracket on the spring seat plate.

e. **Install Hydraulic Brake Line and Propeller Shaft.** Install the flexible hydraulic line to the connection at the differential housing (fig. 77). Connect the propeller shaft at the axle (par. 21 **a**).

f. **Lubricate.** Fill the differential to proper level with specified oil. Apply specified grease to all fittings. Bleed the hydraulic brake system. Refer to TM 9-803.

855473 O - 49 - 7

Section VII

FITS AND TOLERANCES

35. FITS AND TOLERANCES.

Fits Location and Name	Manufacturers Fit Tolerance	Fit Wear Limit	Type of Fit
a. Transmission.			
Second speed gear bushing....	—	—	Press
Second speed gear and mainshaft	0.001-0.002 in.	0.004 in.	Running
Idle gear bushing	—	—	Press
Idle gear and idle gear shaft	0.003-0.0045 in.	0.005 in.	Running
Countershaft end play	0.004-0.016 in.	0.016 in.	—
Countershaft gear bushings and countershaft gear	0.0015-0.003 in.	0.005 in.	Running
Countershaft gear bushings and countershaft	0.0015-0.0025 in.	0.005 in.	Running
b. Transfer Case.			
Intermediate gear end play	0.006-0.017 in.	0.017 in.	—
Output shaft bushing and clutch shaft	0.0015-0.003 in.	0.003 in.	Running
Shift lever pivot pin and shift levers	0.001-0.005 in.	0.010 in.	Slip
Output shaft and output shaft gear	0.0015-0.0025 in.	0.003 in.	Running
c. Front Axle.			
Differential pinion gears and differential pinion shaft	0.0019-0.0044 in.	0.005 in.	Running
Axle shaft gear and differential case	0.003-0.006 in.	0.006 in.	Running
Differential pinion adjustment	0.719 in.	0.719 in.	—
Differential ring gear backlash	0.005-0.007 in.	0.005-0.007 in.	—
Differential ring gear run-out	0.003 in.	0.003 in.	—
Spindle housing tension	4 to 6 lb	4 to 6 lb pull scale	—
Bendix or Tracta axle shaft backlash	0.015-0.035 in.	0.015-0.035 in.	—
d. Rear Axle.			
Differential pinion gears and differential pinion shaft	0.0019-0.004 in.	0.005 in.	Running
Axle shaft gear and differential case	0.003-0.006 in.	0.006 in.	Running
Differential ring gear backlash	0.005-0.007 in.	0.005-0.007 in.	—
Differential pinion adjustment	0.719 in.	0.719 in.	—
Differential ring gear run-out	0.003 in.	0.003 in.	—

CHAPTER 3

BODY AND FRAME

Section I

SPRINGS AND SHOCK ABSORBERS

36. **SPRINGS.**

a. **Description and Data.**

(1) **Description. The** front and rear springs are the semi-elliptic type. The front end of the front springs and the rear end of the rear springs are shackled, using the U-bolt type shackle with a threaded core bushing. The rear ends of the front springs and the front ends of the rear springs each have a bronze bushing and are each pivoted by a pivot bolt mounted to a bracket on the frame. A torque reaction spring, mounted on the left front spring, stabilizes the torque of the front axle. The front springs appear to be identical in construction but are different in load carrying ability. The left spring can be identified by the letter *"L"* stamped on the No. 8 leaf.

(2) DATA.

Front spring:

Make	Mather
Type leaf	Parabolic
Length (center to center of eye)	36¼ in.
Width	1¾ in.
Number of leaves	8
Front eye (center to center bolt)	181/s in.

Rear eye (center to center bolt) 181/a in.
Left camber under 525 lb •~ in.
Right camber under 390 lb 6 in. Rear eye Bushing

Rebound clips	*4*

Rear springs:

Make	Mather
Type leaf	Parabolic
Length	42 in.
Width	**1¾ in.**
Number of leaves	**9**
Rebound clips	**4**

Camber under **800** lb ¾ in.
Eye to center bolt 21 **in.** Front eye Bushing

Figure 83 — Left Front Spring With Torque Reaction Spring

b. Removal.

(1) RIGHT FRONT SPRINGS (fig. 35). Raise the vehicle frame until the weight is off the springs but the wheels are still on the floor. Remove the cotter pin and flat washer that secure the shock absorber to the spring seat plates. Remove the shock absorbers from the spring seat plates. Remove the four nuts from the spring U-bolts and remove the U-bolts and spring seat plates. Remove the two front shackle bushings from the spring shackles at the forward end of the frame. Remove the cotter pin and nut from the shackle bolt at the rear of the spring. Remove the shackle bolt from the spring. Remove the spring from the vehicle.

(2) LEFT FRONT SPRING (fig. 83). Raise the vehicle frame until the weight is off the springs but the wheels are still on the floor. Remove the cap screw that secures the shackle bolt lock plate to the left side of the frame. Remove the nut and bolt from the clamping end of the lock plate and remove the lock plate from the shackle bolt. Remove the cotter pin and flat washer that secure the lower end of the shock absorber to the torque reaction spring. Pull the shock absorber off the reaction spring. Remove the cotter pin and nut from the reaction spring shackle bolt and remove the shackle bolt. Remove the cotter pin and nut from the spring shackle bolt and remove

the shackle bolt and shackles from the spring. Remove the four nuts from the U-bolts and remove the torque reaction spring. Remove the two spring shackle bushings from the spring shackle at the forward end of the spring. Remove the spring from the vehicle.

(3) REAR SPRINGS (fig. 77). Raise the rear of the vehicle frame until the weight is off the spring but the wheels still are on the floor. Remove the cotter pin and flat washer that secure each shock absorber to the spring seat plate. Remove the shock absorbers from the spring seat plates. Remove the four nuts from the spring U-bolts at both springs. Remove the U-bolts and spring seat plates. Remove the two shackle bushings from the spring shackle at the rear of the spring. Remove the spring shackles from the spring. Remove the cotter pin and castellated nut from the two shackle bolts at the front of the rear spring. Remove the two shackle bolts from the springs. Remove the rear springs from the vehicle.

c. Cleaning, Inspection, and Repair.

(1) CLEANING AND INSPECTION (figs. 85 and 86). Clean all parts in dry-cleaning solvent. Replace spring leaves or spring clips that are cracked or bent (step (2) *(b)*, below). Replace spring shackles or shackle bolts that are bent or excessively worn. Replace the shackle bolt if the diameter is worn to less than 0.055 inch. Replace the spring bushing in the spring if the inside diameter is worn to more than 0.565 inch (step (2) *(a)*, below). Replace the torque reaction leaves if they are cracked or bent. Replace the bushing in the torque reaction spring if worn to more than 0.566 inch (step (2) *(a)*, below). Replace the inner shackle bushing if the inside diameter is worn to more than 0.570 inch. Replace the outer shackle bushing if the inside diameter is worn to more than 0.630 inch.

(2) REPAIR.

(a) Front and Rear Spring and Torque Reaction Spring Bushing Replacement (fig. 84). Place the spring in a press and, with a suitable driver, press out the bushing. Press a new bushing in the spring, using the same driver.

(b) Spring Leaf Replacement (fig. 86). Remove the nut and bolt from each of the four spring clips and remove the clips. Install a C-clamp next to the spring tie bolt to hold the tension of the spring leaves before removing the tie bolt. Remove the nut from the spring tie bolt and remove the spring tie bolt from the spring. Remove the C-clamp and separate the spring leaves. Replace the damaged or broken spring leaves. To reassemble the spring, place the spring leaves on the spring tie bolt, starting with the shortest leaf. Pull the leaves together in a vise or a suitable press and install the nut on the tie bolt. Install the four spring leaf clips on the spring.

Figure 84 — Pressing Bushing Out of Spring

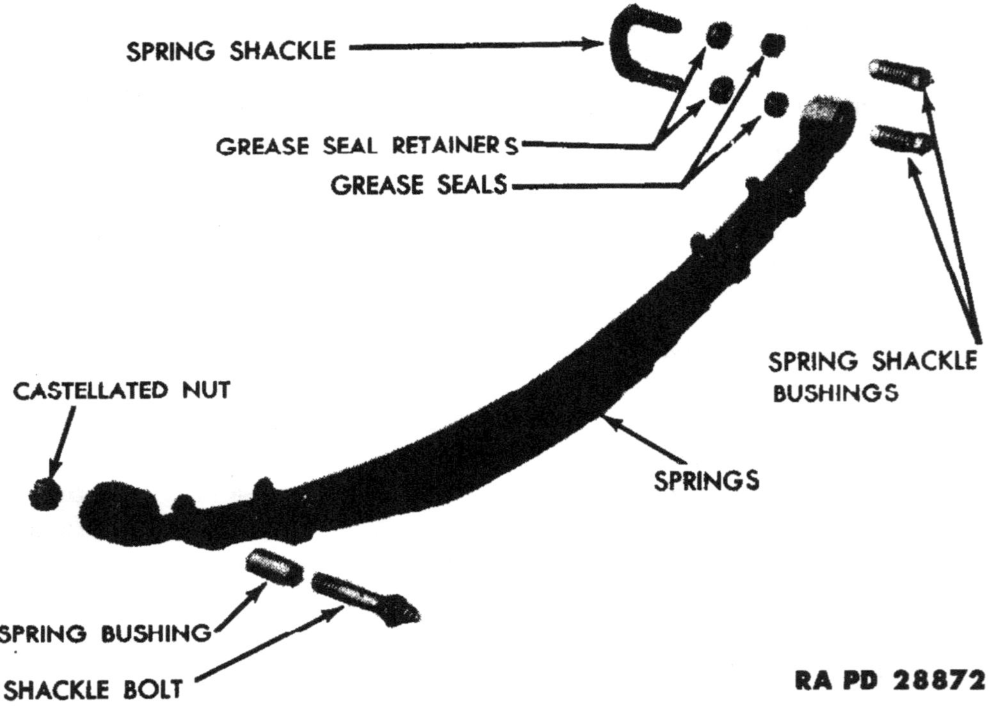

Figure 85 — Rear Spring and Shackles

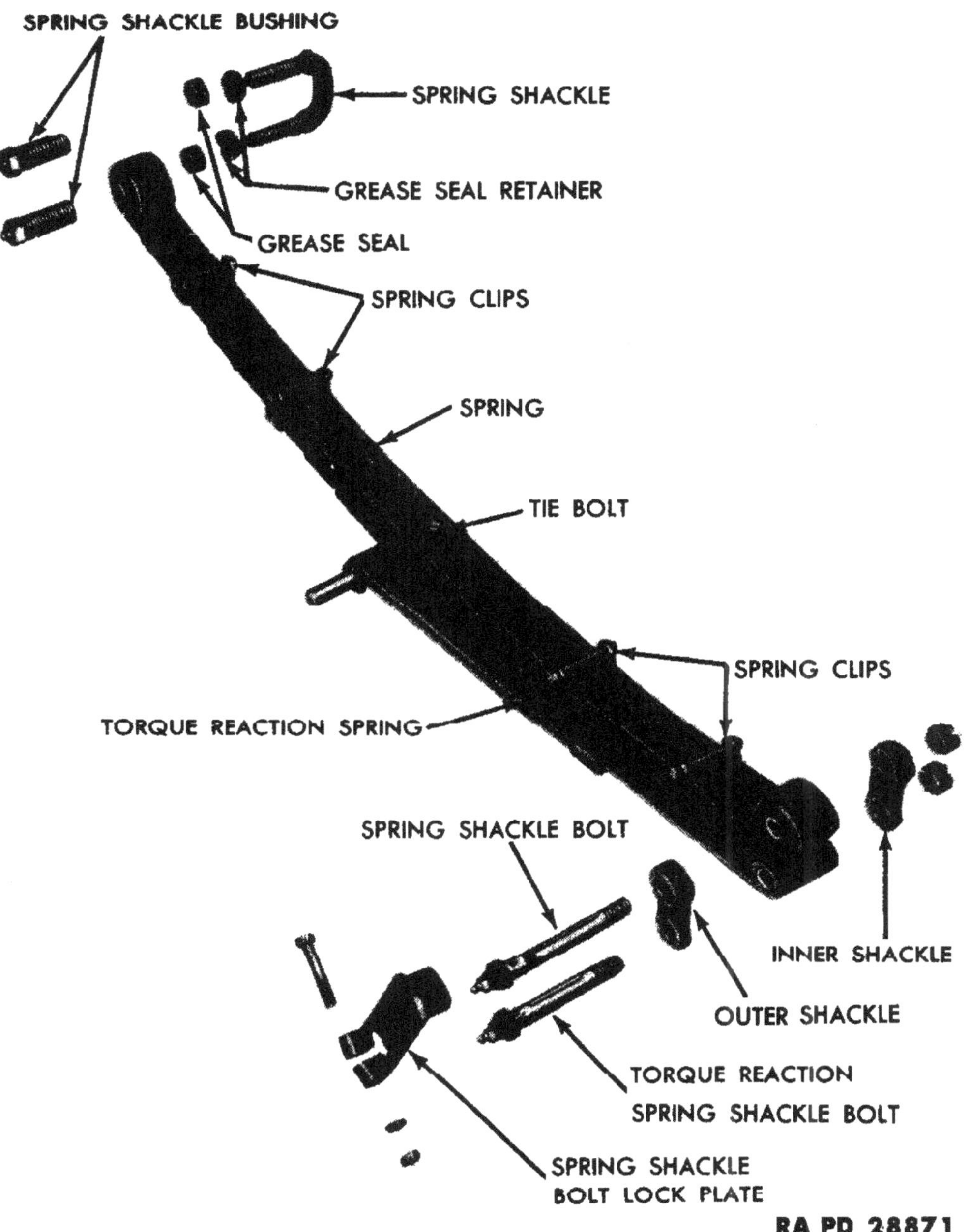

Figure 86 — Front Spring — Exploded View

d. Installation.

(1) RIGHT FRONT SPRING (fig. 35). Place the front spring with the bushing end in the spring bracket on the frame. Insert the spring shackle bolt in the spring with the grease fitting facing outward. Install the nut and cotter pin on the shackle bolt. Raise the forward end of the spring and insert the spring shackle in the bracket on the frame and in the spring. Install the spring shackle bushing with the grease fittings facing outward. Place the spring U-bolts in position

on the axle. Place the spring seat plate on the U-bolts. Install the the four nuts that secure the U-bolts to the axle housing. Install the lower end of the shock absorbers to the spring seat plate. Apply specified lubricant to all fittings.

(2) LEFT FRONT SPRING (fig. 83). Place the spring with the bushing end in the spring bracket on the frame. Insert the outer shackle on the shackle bolt. Insert the shackle bolt in the spring. Place the inner shackle on the shackle bolt and install the nut and cotter pin. Place the torque reaction spring between the inner and outer shackles. Insert the shackle bolt through the shackle and spring. Install the nut and cotter pin on the shackle bolt. Raise the forward end of the spring and insert the spring shackle in the spring. Install the spring shackle bushings with the grease fittings facing outward on the spring shackles. Place the spring U-bolts on the axle housing. Raise the torque reaction spring onto the U-bolts. Install the four nuts to the U-bolts. Install the lower end of the shock absorber to the torque reaction spring. Apply specified lubricant to all fittings.

(3) REAR SPRINGS (fig. 77). Place the rear spring with the bushing end in the spring bracket on the frame. Insert the spring shackle bolt in the spring with the grease fitting facing outward. Raise the rear end of the spring and insert the spring shackle in the spring and in the bracket on the frame. Install the two spring shackle bushings with the grease fitting facing outward. Place the spring U-bolts in position on the axle housing. Place the spring seat plate on the U-bolts. Install the four nuts that secure the spring seat plate to the U-bolts. Insert a rubber mounting in each side of each shock absorber eye. Place the lower end of the shock absorber on the bracket at the spring seat plate. If new shock absorber rubber mountings are being used, compress them with compressor 41-C-2554-400. Install the flat washer and cotter pin that secure the shock absorber to the spring seat plate.

37. GABRIEL SHOCK ABSORBER.

a. Description and Data.

(1) DESCRIPTION. The Gabriel shock absorbers used on some of the vehicles can be distinguished from the Monroe type (par. 38) in that the upper tube has no cutaway section (fig. 87). Four of these direct-acting shock absorbers are used, one at each side of each axle. These shock absorbers are sealed at the factory with the proper amount of fluid and are non-refillable. These shock absorbers are adjustable (subpar. e, below).

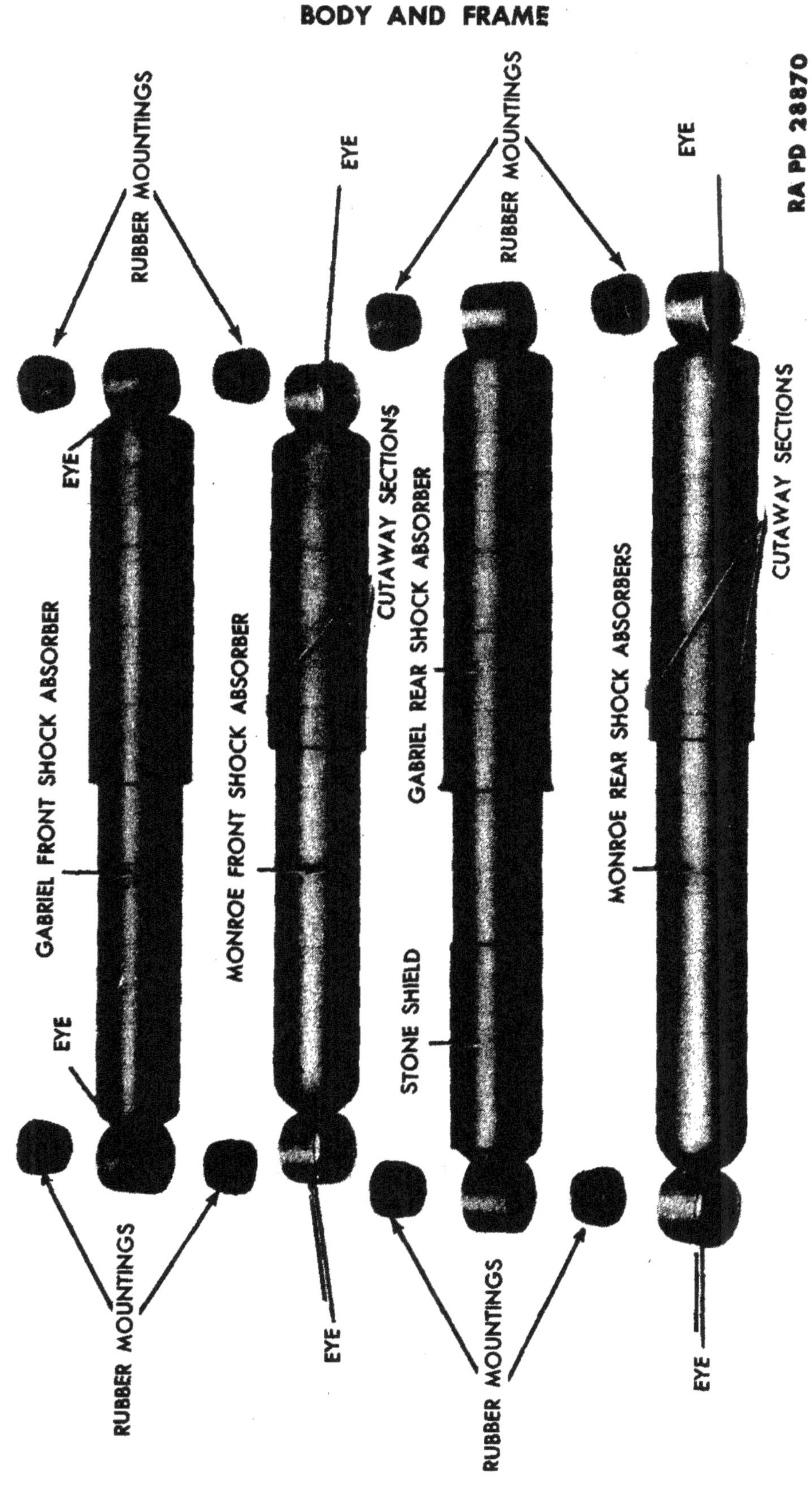

Figure 87 — Monroe and Gabriel Shock Absorbers

(2) DATA.

Make .... Gabriel
Type .... Hydraulic
Action .... Double
Length compressed:
Front .... 10⁵⁄₁₆ in.
Rear .... 11⁵⁄₁₆ in.
Length extended:
Front .... 16⁵⁄₁₆ in.
Rear .... 18⁵⁄₁₆ in.
Mountings .... Rubber

b. Removal (figs. 35 and 77). Remove the cotter pin and flat washer that secure the upper end of the shock absorber to the bracket on the frame. Remove the cotter pin and flat washer that secure the lower end of the shock absorber to the spring seat plate. Remove the shock absorber and rubber mountings from the vehicle.

c. Cleaning and Inspection. Wash the shock absorber with dry-cleaning solvent. If the shock absorber is cracked, excessively worn or is leaking fluid, replace the shock absorber. Replace the rubber mountings if they are excessively worn. Do not clean the shock absorber rubber mountings in dry-cleaning solvent.

d. Installation (figs. 35 and 77). Insert a rubber mounting in each side of the upper and lower eye of the shock absorber. Place the shock absorber onto the spring seat plate and onto the bracket on the frame. Install the flat washer and cotter pin that secure the upper end of the shock absorber to the frame. Install the flat washer and cotter pin that secure the lower end of the shock absorber to the spring seat plate. Install the rear shock absorbers so that the stone shield (fig. 87) on the shock absorber is facing forward on the vehicle.

e. Adjustment. Remove the cotter pin and flat washer from the lower end of the shock absorber and remove the lower end from the bracket. Push the unit together to engage the adjusting key, turn the lower half of the shock absorber clockwise until the limit of the adjustment is reached. Holding the unit together to keep the adjusting key still in the slot, turn the lower end of the shock absorber back (counterclockwise) one-half turn. This is the standard adjustment. Turning the adjustment to the right (clockwise) gives a firmer control for rough terrain, turning the adjustment counterclockwise establishes a softer control.

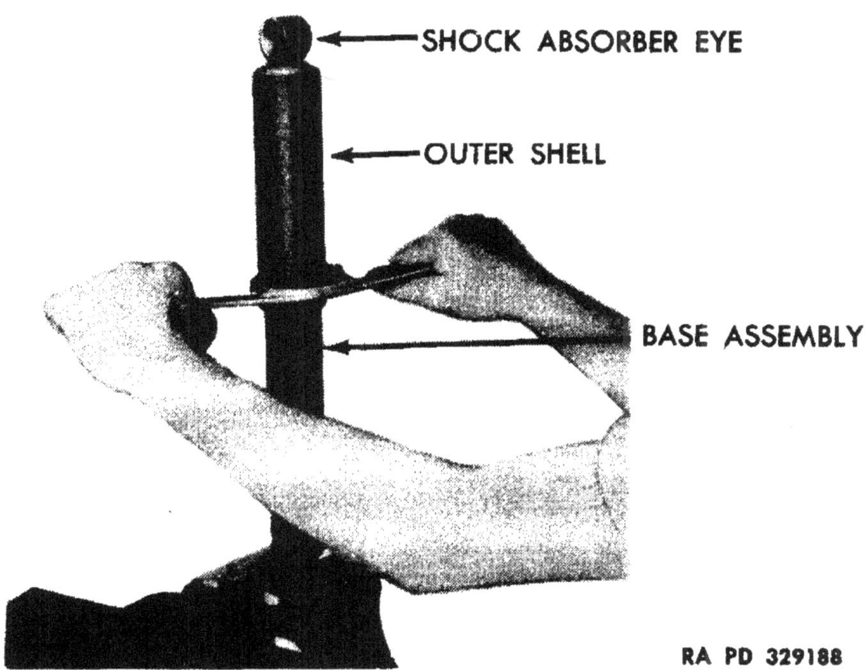

Figure 88 — Removing Seal Assembly With Special Spanner Wrench 41-W-3336-745

38. MONROE SHOCK ABSORBER.

a. Description and Data.

(1) DESCRIPTION. The Monroe type shock absorber used on some of the vehicles can be distinguished from the Gabriel shock absorber by the cutaway sections on the outer shell of the shock absorber (fig. 87). Four of these direct-acting shock absorbers are used, one on each spring. These shock absorbers are refillable (subpar. e (5), below) and can be disassembled for repairs. They are also adjustable (subpar. e (3), below).

(2) DATA.

Make .... Monroe
Type .... Hydraulic
Action .... Double
Length, compressed:
  Front .... 10 9/16 in.
  Rear .... 11 9/16 in.
Length, extended:
  Front .... 16 1/8 in.
  Rear .... 18 1/8 in.
Mountings .... Rubber

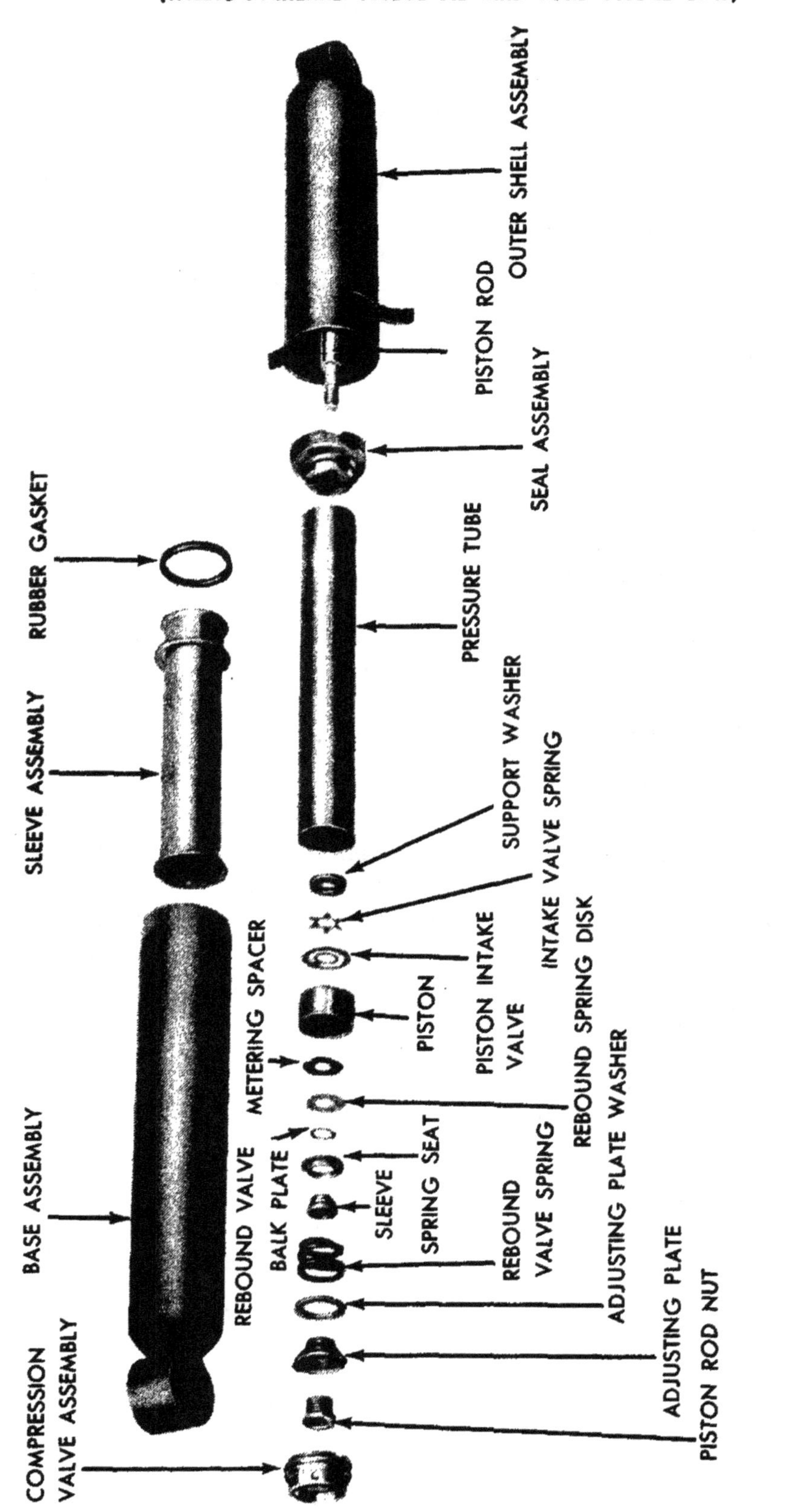

Figure 89 — Monroe Shock Absorber — Exploded View

b. Removal. Remove the cotter pin and flat washer that secure the upper end of the shock absorber to the bracket on the frame. Remove the cotter pin and flat washer that secure the lower end of the shock absorber to the spring seat plate. Remove the shock absorber and rubber mountings from the vehicle. The same procedure applies to all four shock absorbers.

c. Disassembly (figs. 88 and 89). Place the eye of the base assembly in a vise. Pry open the two metal punch-out openings at the lower end of the outer shell. Install the special spanner wrench 41-W-3336-745 in the slots of the seal assembly and unscrew the seal assembly from the base. Pull the outer shell and pressure tube out of the base. Remove the base assembly from the vise and install the eye of the outer shell in the vise. Pry the compression valve assembly off the pressure tube, using a pair of pliers. Remove the shock absorber from the vise. Turn the pressure tube upside down and remove the fluid. Place the eye of the outer shell back in the vise in its original position. Push the pressure tube down into the outer shell. Remove the piston rod nut. Pull the pressure tube off the piston rod. Remove all of the internal parts from the pressure tube. Place a long drift in the pressure tube and tap the seal assembly out of the pressure tube.

d. Cleaning and Inspection (fig. 89). Clean all parts in dry-cleaning solvent. Replace the rubber gasket and seal assembly regardless of its condition. Replace all parts that are cracked, bent or excessively worn. Replace the piston if the diameter is worn to less than 0.997 inch. Replace the pressure tube if the inside diameter is worn to more than 1.001 inches. Replace the outer shell if the piston rod is bent or excessively worn. Replace the compression valve assembly if the valve spring is broken or if the adjustment slots are excessively worn. Replace the base assembly if there are any bad dents in the casing or if the threads are damaged. Replace the sleeve assembly if it is bent or out of shape.

e. Assembly.

(1) INSTALL PRESSURE TUBE AND INTERNAL PARTS (fig. 89). Place the eye of the outer shell in a vise. Place the seal assembly at either end of the pressure tube and press the seal assembly in the tube. Install the special thimble 41-T-1657 on the threaded end of the piston rod. Push the pressure tube down into the outer shell and remove the pilot tool from the piston rod. Place the following parts on the piston rod in the order given; piston support washer with the flat surface facing down; intake valve spring with the bent ends facing up; piston intake valve; piston with the skirt of the piston facing up; metering spacer; rebound spring disk; rebound valve back plate; spring seat with the flat surface facing down; sleeve with the tapered

end facing down; rebound valve spring; and adjusting plate washer. Screw the piston rod nut all the way into the adjusting plate. Install the nut and adjusting plate on the piston rod. Stake the rod and nut to prevent the nut from loosening.

(2) FILL SHOCK ABSORBER WITH FLUID. Pull the pressure tube out of the outer shell to its fullest extent. If working on a front shock absorber, measure 5 ounces of shock absorber fluid and put it in a clean container or 5¾ ounces if working on a rear shock absorber. Fill the pressure tube with fluid from this container to within ⅜ inch from the top. Pour the remaining amount of the measured fluid into the base assembly. Hold the pressure tube firmly and place the compression valve assembly on the tube. Tap the valve lightly until it is seated firmly in the pressure tube. Remove the outer shell from the vise and install the loop end of the base in the vise. Insert the sleeve assembly into the base assembly. Insert a new rubber gasket into the base assembly. Place the outer shell on the base. Using the special spanner wrench 41-W-3336-745, tighten the seal assembly into the base assembly (fig. 88).

(3) ADJUST. Push the unit together to engage the adjusting key, turn the base assembly (lower half) of the shock absorber clockwise until the limit of the adjustment is reached. Holding the unit together to keep the adjusting key still in the slot, turn the lower end of the shock absorber back (counterclockwise) two turns. This establishes the standard adjustment. Turning the adjustment to the right (clockwise) gives a firmer control for rough terrain, turning the adjustment counterclockwise establishes a softer control.

(4) INSTALL. Insert the rubber mountings in the upper and lower ends of the shock absorber. Install the shock absorber to the spring seat plate and to the frame. Install the flat washer and cotter pin that secure the upper end of the shock absorber to the frame. Install the flat washer and cotter pin that secure the lower end of the shock absorber to the spring seat plate.

(5) REFILL SHOCK ABSORBER. Remove the shock absorber (subpar. b, above). Place the eye end of the shock absorber base assembly in a vise. Pry open the two metal punch-out openings at the lower end of the outer shell. Install the special spanner wrench 41-W-3336-745 in the slots of the seal assembly (fig. 88). Unscrew the seal assembly from the base assembly. Pull the outer shell with the pressure tube out of the base assembly. Remove the base assembly from the vise and install the eye end of the outer shell in the vise. Pry the compression valve assembly off the pressure tube, using a pair of pliers. Fill the shock absorber with fluid as outlined in step (2), above.

Section II

STEERING GEAR AND DRAG LINK

39. STEERING GEAR ASSEMBLY.

a. Description. The Ross Model T-12 steering gear (fig. 90) is of the cam and twin lever, variable ratio type, having a ratio of 14-12-14 to 1. The steering gear sector shaft is serrated for attachment of the Pitman arm, and the steering wheel is serrated for attachment to the worm and shaft assembly. The steering wheel is of the safety type, having three spokes and is 17¼ inches in diameter.

b. Removal.

(1) REMOVE LEFT FRONT FENDER. Remove the 12 bolts that secure the left front fender to the body, frame and radiator guard. Remove the bolt that secures the fender to the top of the frame in the engine compartment. Remove the wing nut that secures the headlight bracket to the fender. Disconnect the wires leading from the fender to the junction block on the cowl. Disconnect the wires leading from the junction block on the fender to the headlight and blackout light. Remove the fender from the vehicle.

(2) REMOVE STEERING WHEEL. Remove the steering wheel nut, horn button nut and horn button. Pull the steering wheel off the shaft with a steering wheel puller.

(3) REMOVE STEERING COLUMN TUBE AND BEARING ASSEMBLY. Remove the two nuts and bolts that secure the steering column support clamp at the instrument panel and remove the clamp. Remove the four metal screws that hold the steering column cover plate to the floor plate in the driver's compartment. Remove the two screws that hold the horn wire contact brush to the steering column and remove the brush. Loosen the bolt at the steering column clamp and slide the steering column tube and bearing assembly off the shaft.

(4) DISCONNECT DRAG LINK AT PITMAN ARM (fig. 104). Remove the cotter pin from the Pitman arm end of the drag link. Loosen the drag link socket plug and lift the drag link off the Pitman arm.

(5) REMOVE STEERING GEAR (fig. 104). Remove the three bolts that hold the steering gear to the frame. Slide the steering gear assembly down through the floorboard and out over the frame.

c. Disassembly.

(1) REMOVE PITMAN ARM (fig. 92). Remove the nut and lock washer that hold the Pitman arm on the sector shaft assembly. Pull the Pitman arm off the steering sector shaft assembly with a standard Pitman arm puller.

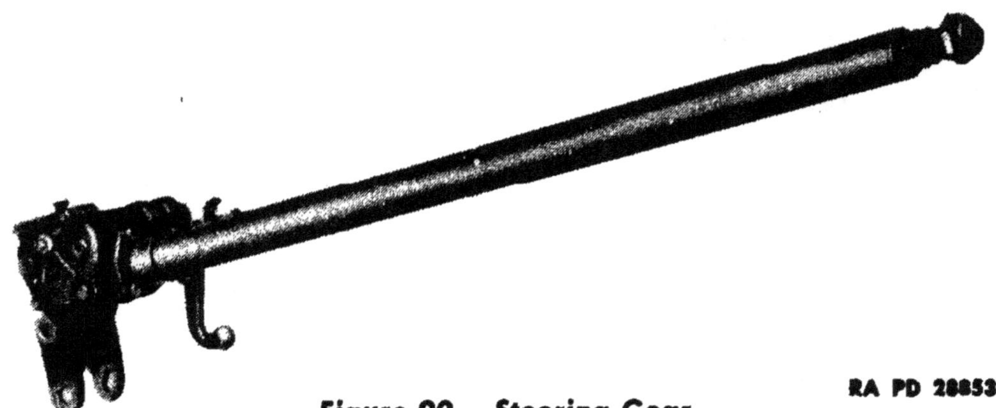

RA PD 28853

Figure 90 — Steering Gear

(2) REMOVE STEERING SECTOR SHAFT ASSEMBLY (fig. 92). Remove the four cap screws that hold the side cover to the housing and remove the side cover and gasket. Slide the sector shaft assembly from the housing.

(3) REMOVE STEERING GEAR WORM AND SHAFT ASSEMBLY FROM HOUSING (fig. 92). Remove the three cap screws that secure the housing end cover and shims to the housing. Slide the housing and shims off the worm and shaft assembly.

(4) REMOVE STEERING GEAR WORM BEARINGS (fig. 91). Remove the retainer ring that secures the steering gear worm lower bearing cup at the end of the shaft assembly. Remove the bearing cup and balls. Remove the retainer ring that secures the worm upper bearing cup to the shaft assembly. Slide the worm bearing cup up on the shaft and remove the balls.

d. Cleaning, Inspection, and Repair.

(1) CLEANING AND INSPECTION (figs. 91 and 97). Clean all parts thoroughly in dry-cleaning solvent. Replace a housing assembly or side cover that is cracked or damaged. Replace the expansion plug in the lower end of the housing if it is loose. Replace the inner and outer bushings in the housing (step (2) *(c)*, below) if worn larger than 0.876 inch inside diameter. Replace a sector shaft assembly that has flat spots on the tapered studs or that has chipped studs. Replace the sector shaft if the shaft measures less than 0.870 inch at the bearing surfaces. Replace the worm and shaft assembly if the worm is excessively worn, ridged, scored, or chipped. Replace a worn, pitted, or cracked worm upper and lower bearing cup (step (2) *(b)*, below). Replace a broken or damaged horn wire (step (2) *(a)*, below). Replace a steering column tube that is bent or damaged. Replace the whole assembly if it is damaged. Replace

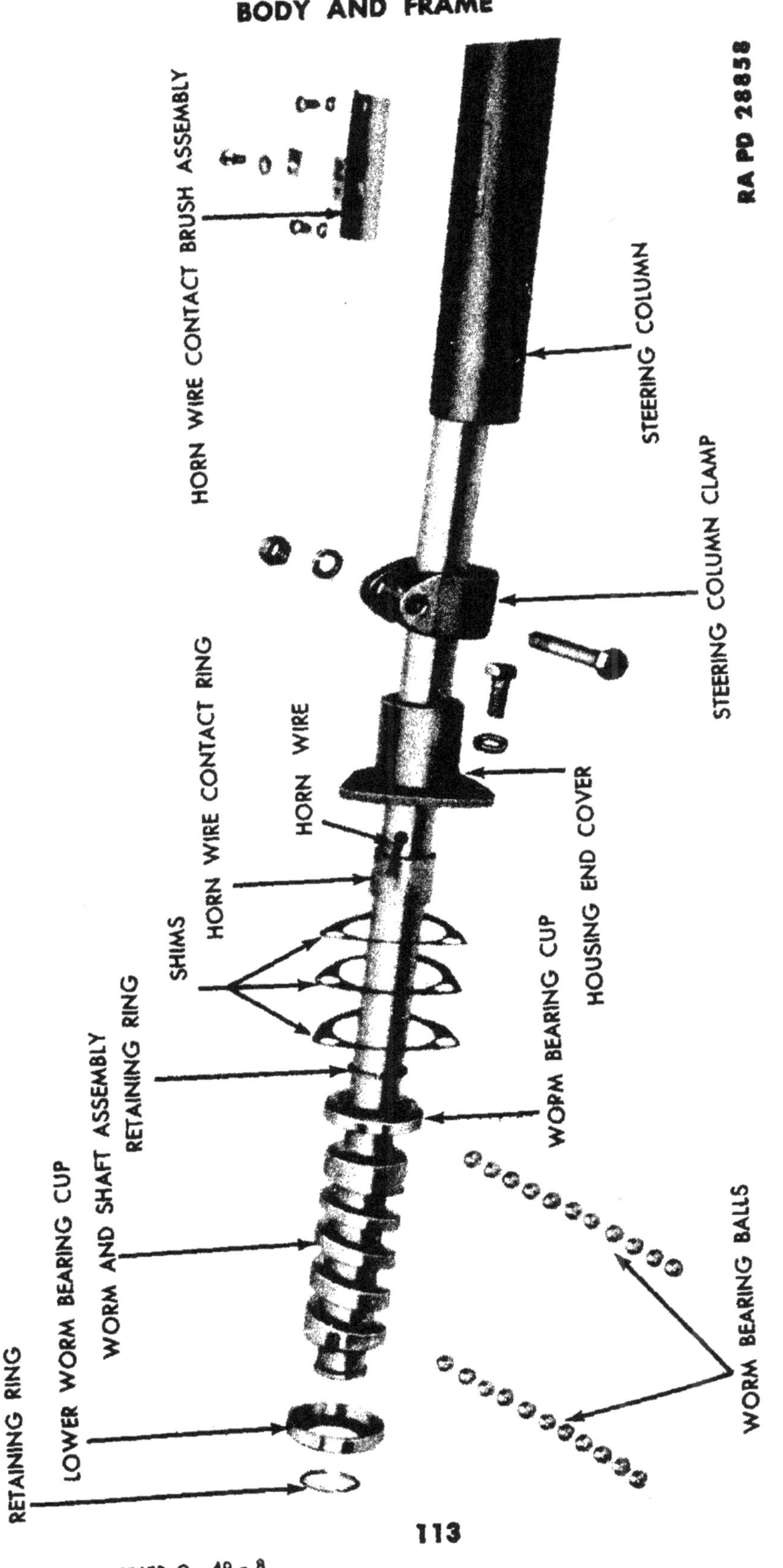

Figure 91 — Worm and Shaft Assembly — Exploded View

855473 O - 49 - 8

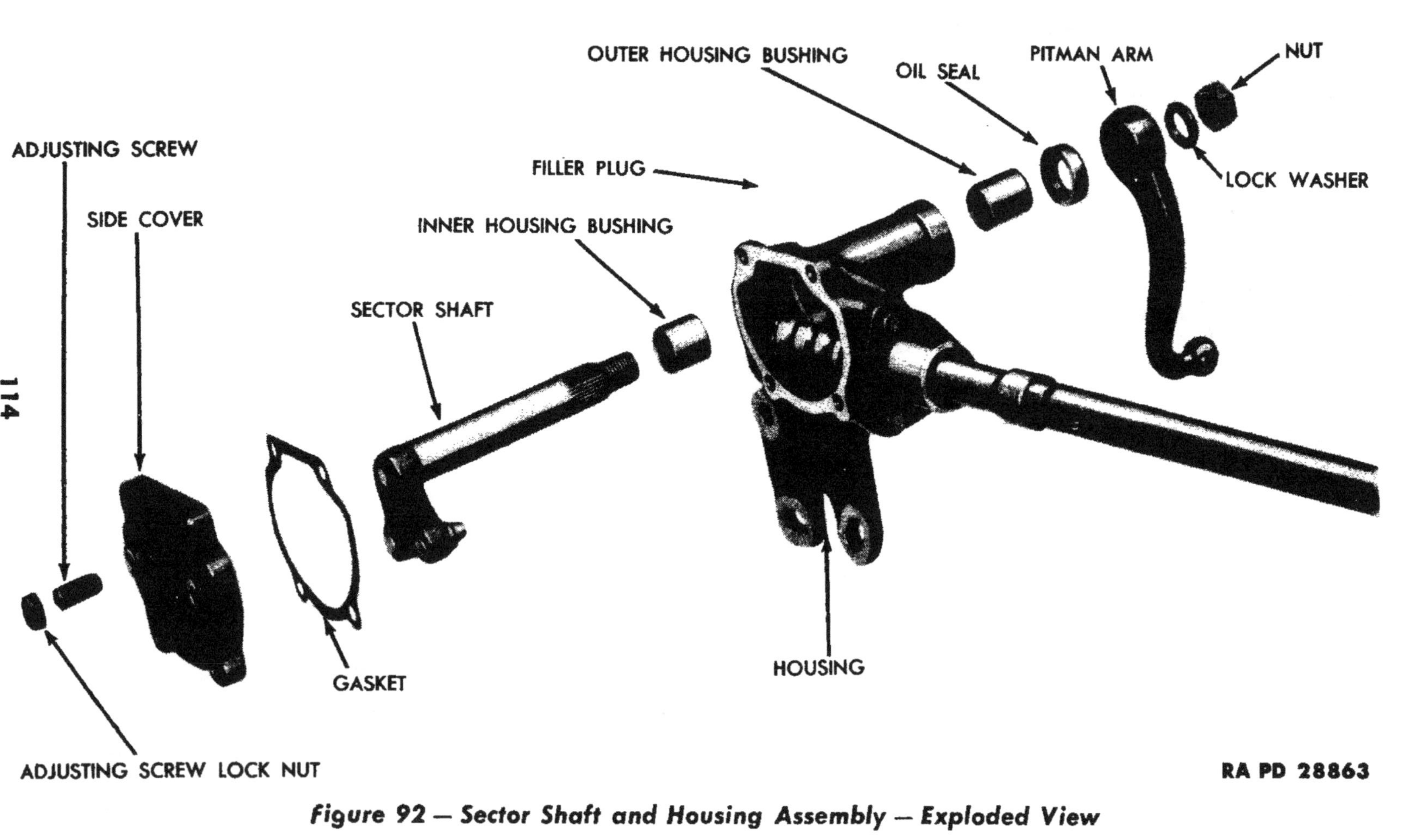

Figure 92 — Sector Shaft and Housing Assembly — Exploded View

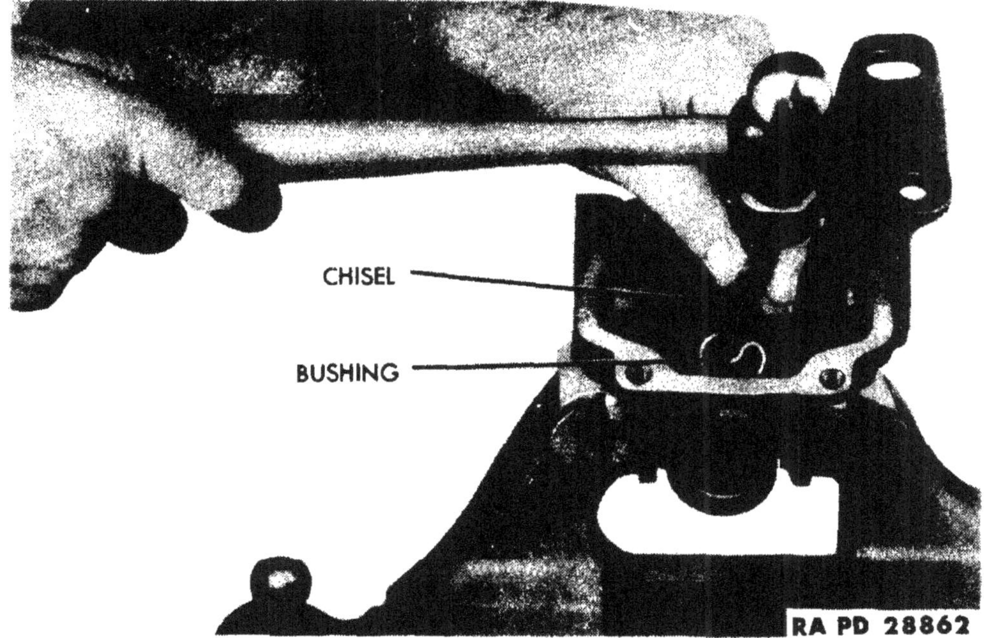

Figure 93 — Removing Housing Bushings

any balls with flat spots. Replace the steering column tube bearing if it is excessively worn.

(2) REPAIR.

(a) Horn Wire Replacement (fig. 96). Unsolder the horn wire at the horn wire contact ring and pull the wire assembly from the shaft. To install the horn wire, slide the contact washer, insulating ferrule, horn button spring and horn button spring cup onto the horn wire. Push the horn wire down through the shaft and out of the hole in the lower end of the shaft. Solder the horn wire onto the horn wire contact ring, making sure that none of the strands of wire or the solder are touching the shaft.

(b) Worm Upper Bearing Cup Replacement. Remove the horn wire (step *(a)*, above). Unsolder the horn wire from the contact ring (fig. 96) and slide the ring off the shaft. Slide the worm upper bearing cap and retaining ring off the shaft. To reinstall the worm bearing cup on the shaft, slide the cup onto the shaft with the concave side toward the worm. Slide the retaining ring onto the shaft. Select a heavy flat washer that will slide onto the shaft, but will not slide over the horn wire contact ring. Make a mark on the lower part of the shaft with a file ½ inch below the horn wire hole. Slide the horn wire contact ring onto the shaft. Place the flat washer on the shaft and place the shaft assembly loosely in a vise (fig. 98). Using a fiber block to protect the shaft, drive down on the shaft until the mark on the shaft is even with the upper part of the horn

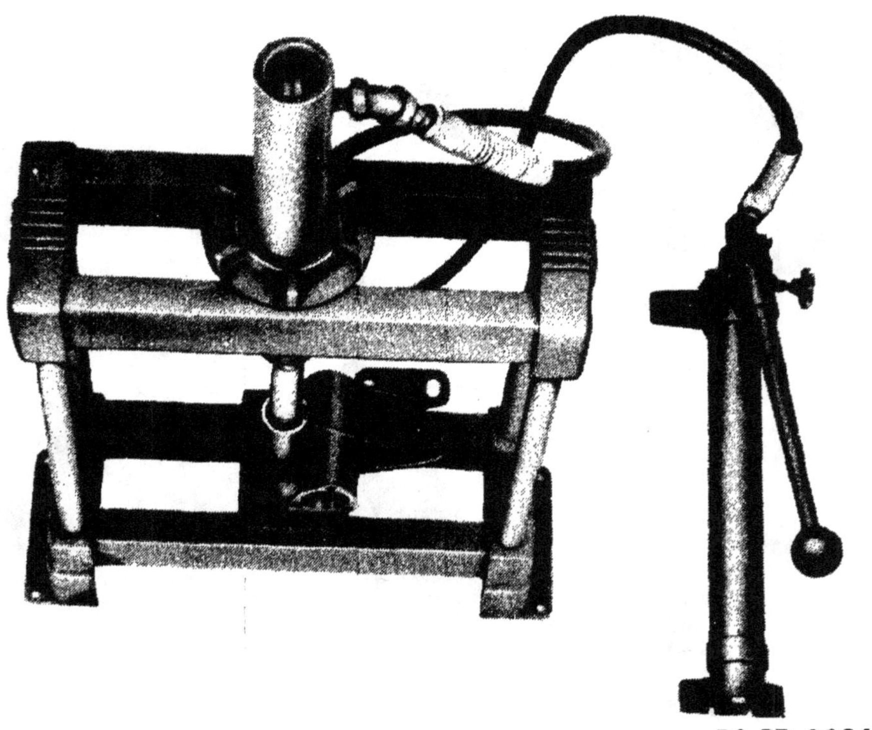

RA PD 28861

Figure 94 — Pressing Bushing in Steering Gear Housing

wire contact ring. Install the horn wire in the shaft (step *(a)*, above).

(c) Housing, Inner, and Outer, Bushing Replacement. Remove the housing oil seal with a small punch or chisel. Drive a small punch or chisel between the housing and the joint in the bushings (fig. 93), until the ends of the bushings overlap. Tap the bushings out of the housing. To install the bushings, press the outer bushing into the housing (fig. 94) until it is flush with the oil seal shoulder in the housing assembly. Press the inner bushing into the housing. Ream the bushings to 0.875 inch diameter (fig. 95) with reamer 41-R-1220.

e. Assembly.

(1) ASSEMBLE WORM AND SHAFT ASSEMBLY. Place the 11 worm bearing balls in the upper worm bearing cup. Slide the cup and retainer ring in place. Place the balls in the lower worm bearing cup and install the cup and retainer ring on the lower end of the shaft.

(2) INSTALL WORM AND SHAFT IN HOUSING. Slide the worm and shaft assembly in the housing. Install shims approximately 0.024 inch thick and the housing end cover on the shaft (fig. 91).

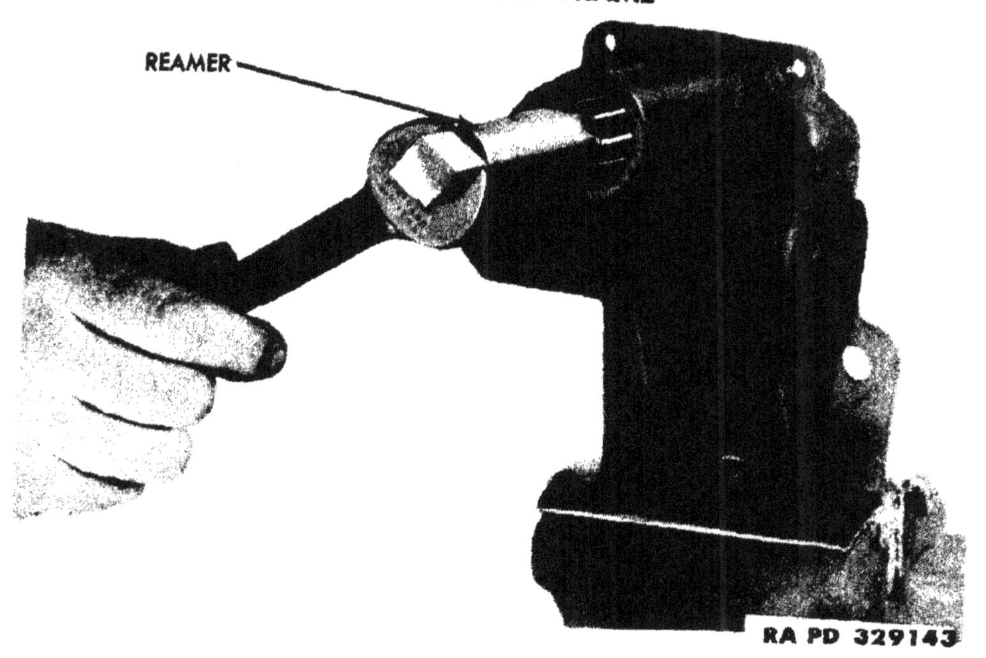

Figure 95 — Reaming Steering Gear Housing Bushing With Reamer 41-R-1220

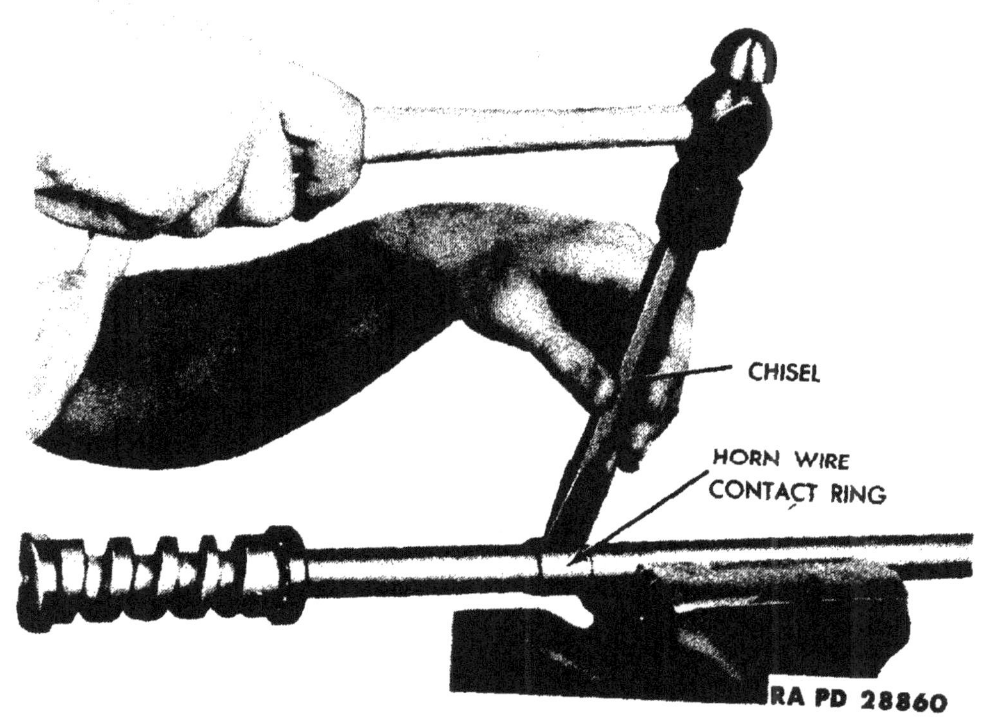

Figure 96 — Removing Horn Wire Contact Ring

Install the three cap screws that secure the housing end cover. Turn the shaft by hand, if it is too tight, shims must be added, if too loose, shims must be removed. The correct adjustment is when shaft turns freely but has no end play. Shims are supplied in 0.0035, 0.0025 and 0.0065 inch thicknesses.

(3) INSTALL SECTOR SHAFT ASSEMBLY (fig. 92). Slide the sector shaft assembly into the housing, making sure the two tapered studs engage in the worm.

(4) INSTALL SIDE COVER ON HOUSING. Install a new side cover gasket and cover on the housing. Install the side adjusting screw and lock nut in the side cover (fig. 92). Turn the shaft counterclockwise until the shaft cannot be turned any farther. Turn the shaft clockwise until it cannot be turned any farther, counting the complete and part revolutions of the shaft and then turn the shaft counterclockwise half the number of turns. This will center the shaft. Turn the side adjusting screw in, until a slight drag on the shaft can be felt at this point. Tighten the adjusting screw lock nut and recheck the adjustment.

(5) INSTALL STEERING COLUMN TUBE ON SHAFT (fig. 91). Install the steering column clamp on the steering column with the bolt side of the clamp in line with the horn wire contact brush opening in the steering column. Slide the steering column tube down onto the housing end cover, making sure the horn wire contact brush opening in the steering column is in a vertical position when the housing base is in normal position. Tighten the clamp bolt.

(6) INSTALL PITMAN ARM (fig. 91). Place the Pitman arm on the sector shaft assembly, making sure the line on the Pitman arm is in line with the mark on the sector shaft assembly and that the ball joint on the Pitman arm is facing downward when the shaft is centered. Install the lock washer and nut that secure the Pitman arm to the sector shaft.

(7) INSTALL HORN WIRE CONTACT BRUSH ASSEMBLY. Hold the horn wire contact brush assembly in place on the steering column and install the two hold-down screws.

f. Installation.

(1) INSTALL STEERING ASSEMBLY IN VEHICLE. Slide the steering gear up through the floorboard. Install, but do not tighten, the three steering gear bolts.

(2) CONNECT STEERING GEAR TO BRACKET. Slide the steering column cover plate down the steering column and fasten it to the floorboard with four metal screws. Install the two bolts that secure the steering column clamp to the instrument panel in the driver's compartment.

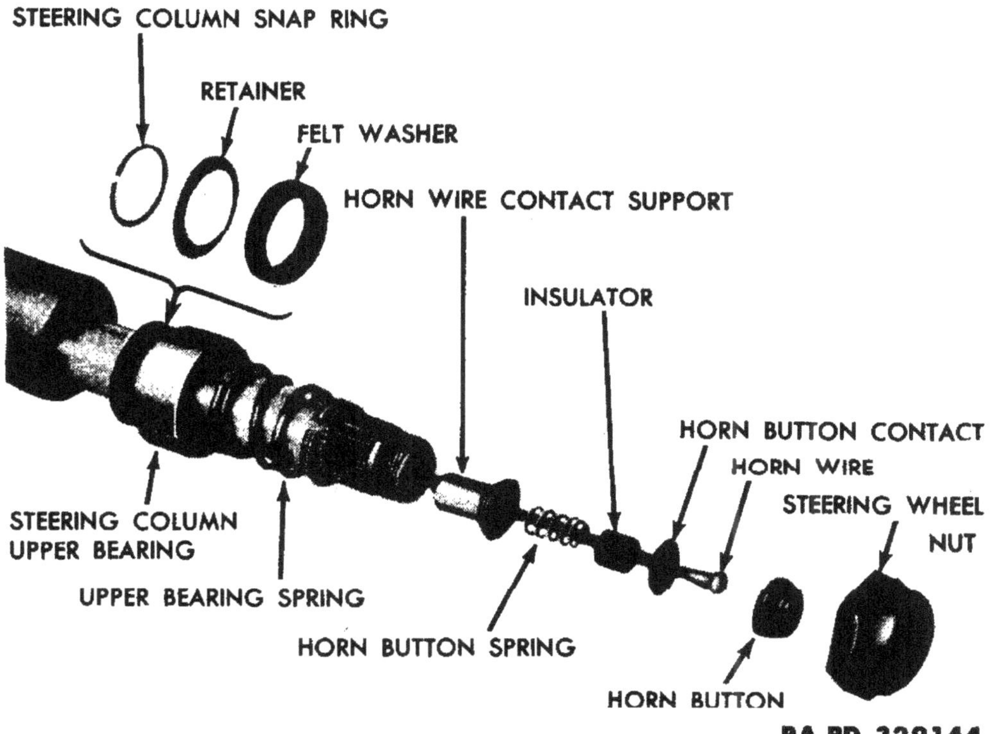

Figure 97 — Horn Button Assembly — Exploded View

(3) CONNECT DRAG LINK (fig. 104). With the front wheels in a straight ahead position, connect the drag link to the Pitman arm, making sure the ball socket on the Pitman arm is between the socket and ball seat (fig. 99). Tighten the socket plug and install the cotter pin. Tighten the three steering gear hold-down bolts at the frame.

(4) INSTALL HORN WIRE CONTACT BRUSH (fig. 91). Hold the horn wire contact brush in place on the steering column and install the two hold-down screws. Connect the horn (live) wire to the contact brush.

(5) INSTALL STEERING WHEEL (fig. 97). Slide the steering column bearing spring and spring seat on the shaft. With the front wheels in a straight ahead position, place the steering wheel on the shaft with the center spoke in a vertical position and facing downward. Install the horn button and horn nut.

(6) INSTALL FENDER AND REPLENISH LUBRICANT. Place the fender in position on the vehicle. Install the 12 bolts that secure the fender to the frame, body, and radiator guard. Install the bolt that secures the fender on the frame in the engine compartment. Install the wires leading from the fender to the junction block on the cowl. Install the wires leading from the junction block on the

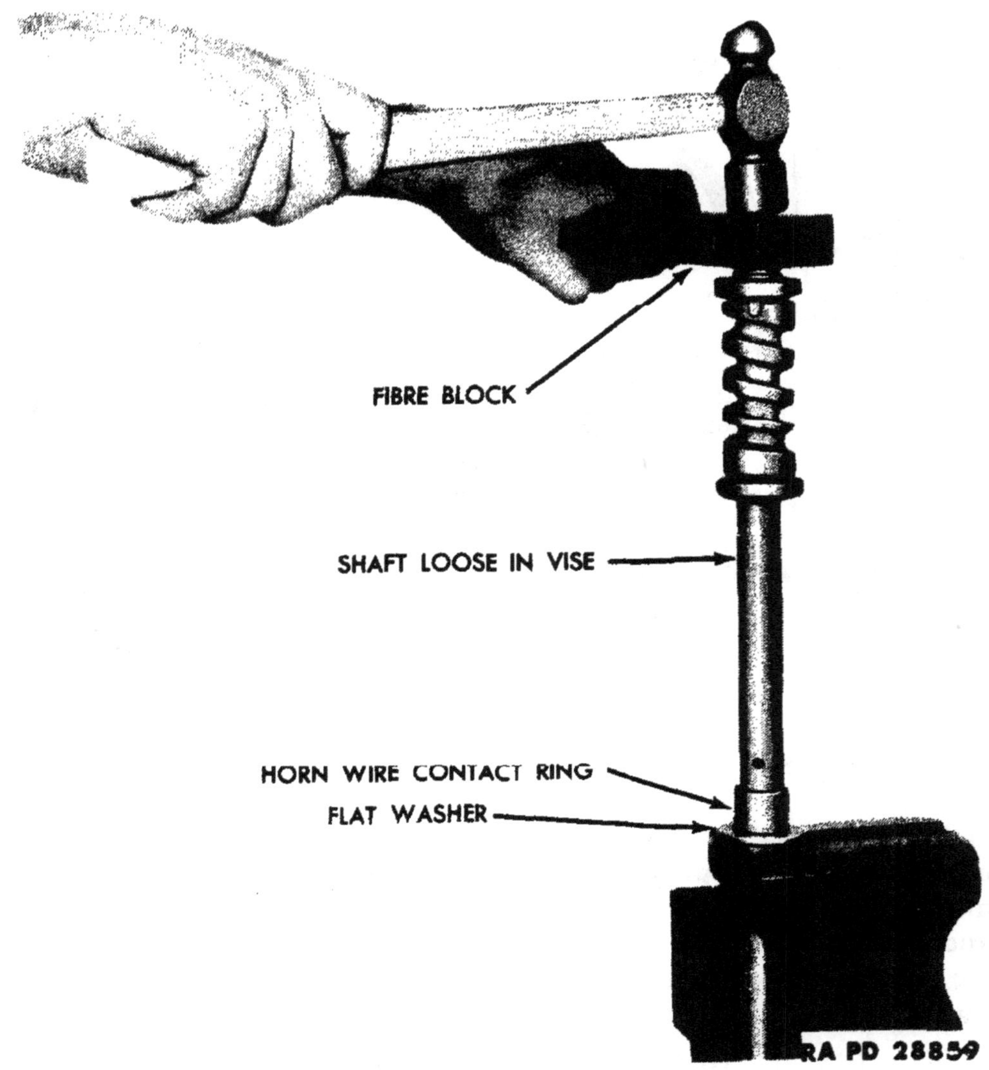

Figure 98 — Installing Horn Wire Contact Ring

fender to the headlight and blackout light. Replenish the lubricant, using the specified amount and grade (refer to TM 9-803).

40. DRAG LINK.

a. **Removal.** Remove the cotter pin from each end of the drag link. Loosen the adjusting plug at each end of the drag link. Lift the drag link from the vehicle.

b. **Disassembly.** Remove the dust seal shield and dust seal from each end of the drag link. Remove the adjusting plug, ball seat, spring, and spring seat from the front end of the drag link. Remove

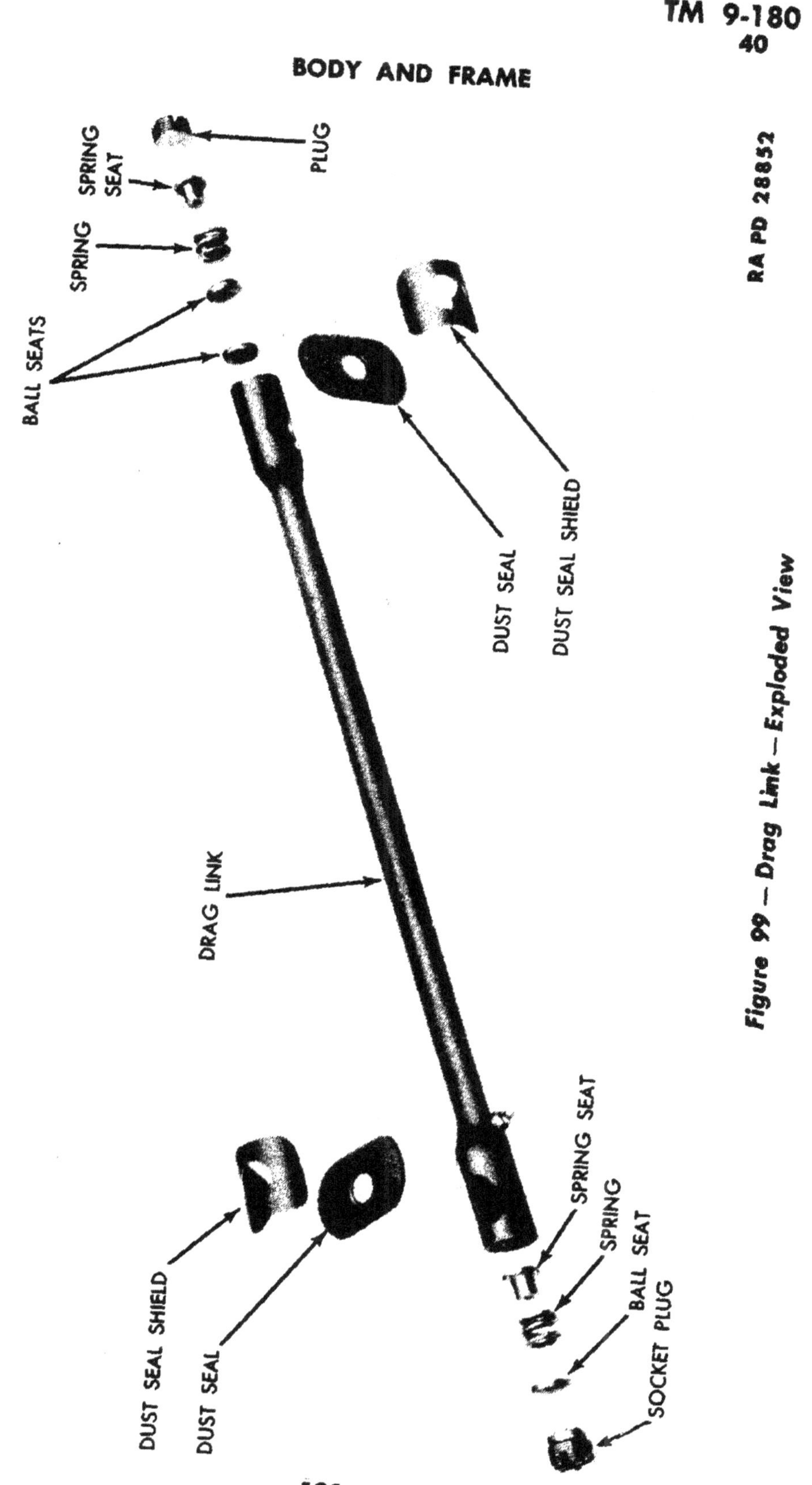

Figure 99 — Drag Link — Exploded View

the adjusting plug, spring seat, spring, and the two ball seats from the rear end of the drag link.

c. **Cleaning, Inspection, and Repair.** Clean all parts thoroughly in dry-cleaning solvent. Replace or straighten the drag link if it is bent. Replace damaged grease fittings. With a small wire, clean all grease fittings that are clogged. Replace excessively worn adjusting socket plugs or broken springs. Replace ball seats that are excessively worn. Replace excessively worn spring seats. Replace damaged dust shields or seals.

d. **Assembly** (fig. 99). Place a spring seat, spring and ball seat in the front end of the drag link. Screw the socket plug approximately three or four turns in the front end of the drag link. Place two ball seats, spring and spring seat in the rear end of the drag link. Screw the socket plug approximately three or four turns in the drag link.

e. **Installation** (fig. 99). Hold a new dust seal and a dust seal shield in place on the rear end of the drag link. Place the drag link on the Pitman arm, making sure the ball joint on the Pitman arm is seated between the spring seat and socket plug. Screw the socket plug in firmly against the ball, then back the plug off one full turn and install a cotter pin. Hold a new dust seal and dust seal shield in place on the front end of the drag link. Place the drag link on the intermediate steering arm, making sure the ball joint is seated between the spring seat and socket plug (fig. 99). Screw the adjusting plug in firmly against the ball and back off the socket plug one-half turn and install a cotter pin.

Section III

BODY

41. REMOVAL.

a. **Remove Hood and Windshield.** Raise the hood and remove the five bolts that secure the hood to the cowl. Remove the hood from the vehicle. Remove the wing nuts that secure the windshield at each side of the cowl. Remove the windshield from the vehicle.

b. **Remove Body Bolts From Frame** (fig. 100). Remove the five bolts under each fender that secure the fender to the body. Remove the four bolts that secure the body to the rear crossmember of the frame. Remove the two bolts that secure the body to the pintle hook brace. Remove the two bolts at each side of the frame that secure the body to the rear body brackets on the frame side member. Remove the bolts that secure the body at each side of the

A REAR CROSSMEMBER
B PINTLE HOOK BRACE
C REAR PROPELLER SHAFT
D BODY BRACKETS
E GROUND STRAP
F SPEEDOMETER CABLE
G MUFFLER
H FRONT FENDER
J FRONT PROPELLER SHAFT
K EXHAUST PIPE
L FRONT FENDER
M TRANSMISSION SUPPORT CROSSMEMBER
N TRANSMISSION SHIELD
O HAND BRAKE CABLE
P HAND BRAKE SPRING
Q BODY BRACKETS

RA PD 329148

Figure 100 — Under Side of Body

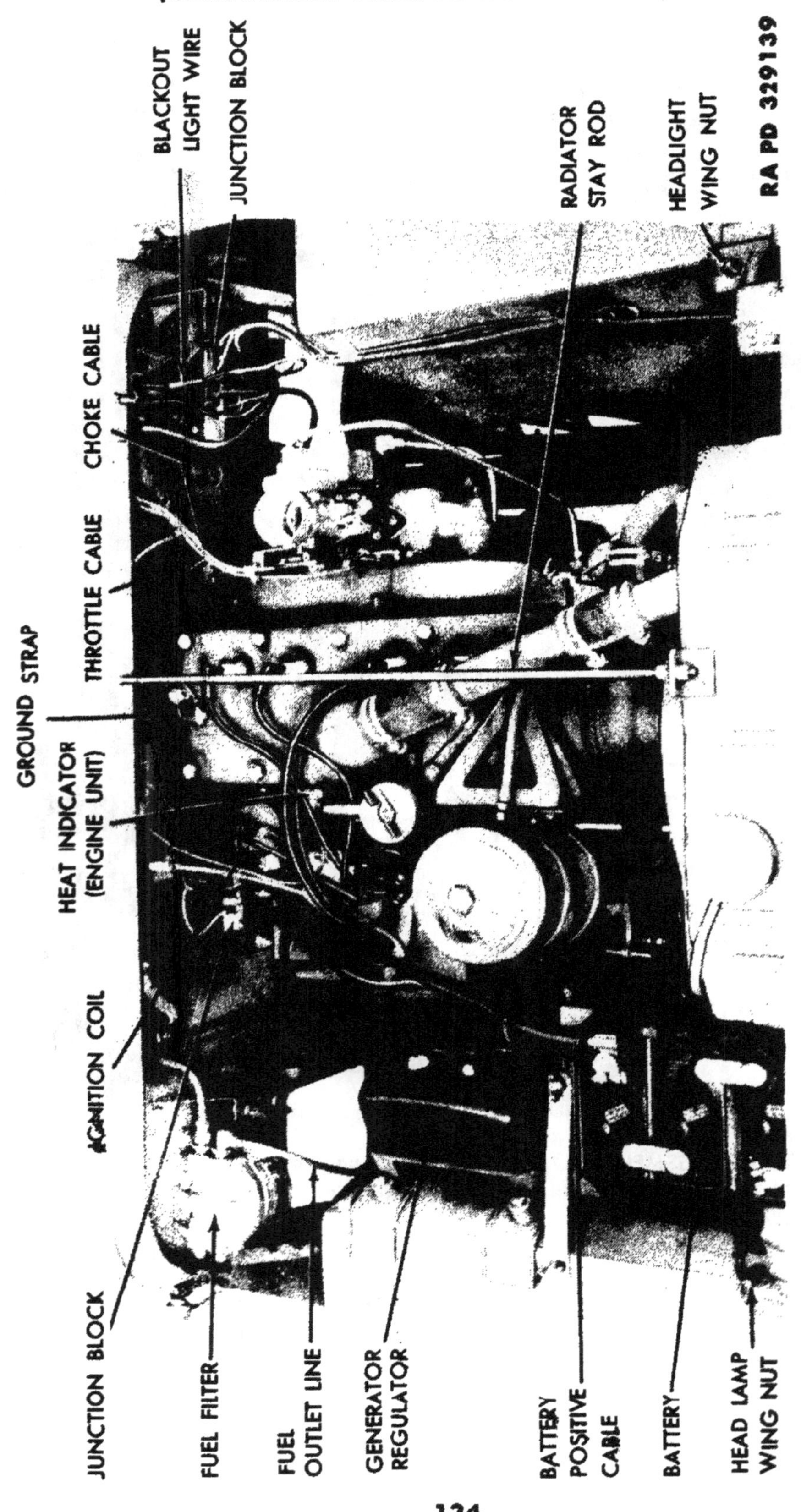

Figure 101 — View of Electrical Wires and Tubes on Cowl

transmission support crossmember. Remove the bolts that secure the body at each side of the frame at the forward end of the body.

c. Disconnect Hand Brake Cable and Speedometer Cable (fig. 100). Remove the hand brake spring at the transfer case. Remove the clevis pin that secures the hand brake cable at the brake on the transfer case. Remove the hand brake cable clamp on the transmission. Disconnect the speedometer cable at the transfer case.

d. Disconnect Ground Straps and Muffler (fig. 100). Disconnect the ground strap at the left side of the transmission. Disconnect the ground strap leading from the body to the right side of the frame at the muffler. Remove the two nuts and bolts that secure the muffler to the body.

e. Remove Air Cleaner. Loosen the hose clamp that secures the air cleaner hose at the air cleaner. Remove the hose from the air cleaner. Remove the four wing nuts that secure the air cleaner to the brackets on the cowl.

f. Remove Clutch and Brake Pedal Pads and Steering Wheel. Remove the cap screw from the clamp at the bottom of the clutch pedal under the floor plate and remove the clutch pedal pads. Remove the cap screw from the clamp under the floor plate at the bottom of the brake pedal and remove the brake pedal. Remove the nut that secures the steering wheel to the steering shaft and remove the steering wheel, using a steering wheel puller. Disconnect the foot accelerator in the driver's compartment.

g. Disconnect Miscellaneous Wires and Tubes in Engine Compartment (fig. 101). Disconnect the positive cable at the battery. Disconnect the wire leading from the junction block at the left side of the cowl to the generator regulator. Disconnect the wire leading from the ignition coil to the junction block. Disconnect the battery cable at the starting switch. Disconnect the ground strap leading from the cylinder to the cowl. NOTE: *When removing the wires, tag them for later identification.* Disconnect the fuel outlet line at the fuel filter. Disconnect the three wires leading from the left fender to the junction block at the left side of the cowl. Disconnect the blackout light wire at the connection at the cowl. Disconnect the two wires at the hydraulic brake master cylinder. Disconnect the wire at the bottom of the steering tube. Disconnect the choke and throttle cable at the carburetor. Disconnect the oil line leading to the oil gage at the flexible connection at the left side of the engine. Disconnect the radiator stay rod at the radiator and cowl and remove the stay rod. Drain the coolant from the radiator and remove the heat indicator (engine unit) at the right-hand side of the cylinder head.

RA PD 28856

Figure 102 — Raising Body From Chassis

h. Remove Body From Frame (fig. 102). Install a rope on the two forward lifting handles on the body and raise the body slowly off the frame. While raising the body, move it toward the rear of the vehicle until the steering tube is clear of the body. Remove the body from the vehicle.

42. INSTALLATION.

a. Place Body in Position on Frame (fig. 102). Install a rope on the two lifting handles at the forward end of the body. Position the body over the chassis. Line up the steering post with the hole provided in the floor plate of the body. Lower the body slowly and at the same time roll the chassis under the body so as to follow the angle of the steering post until the body is resting on the frame.

b. Install Body Bolts (fig. 100). Install the two bolts that secure the body to each side of the frame at the forward end of the body. Install the four bolts that secure the body to the rear crossmember of the frame. Install the two bolts that secure the body to the pintle hook brace. Install the two bolts that secure the body to the two brackets on each frame side member. Install the bolt that secures the body to each side of the transmission support crossmember.

c. Connect Hand Brake Cable and Speedometer Cable (fig. 100). Connect the speedometer cable to the transfer case. Install the clevis pin that secures the hand brake cable at the brake linkage on the transfer case. Install the hand brake cable clamp to the transmission. Install the spring leading from the hand brake to the floor plate.

d. Connect Muffler and Ground Straps (fig. 100). Install the two nuts and bolts that secure the muffler to the right-hand side of the body. Connect the ground strap leading from the left side of the transmission to the floor plate. Connect the ground strap leading from the body to the right-hand side of the frame.

e. Install Miscellaneous Wires and Tubes in Engine Compartment (fig. 101). Install the heat indicator (engine unit) at the right-hand side of the cylinder head. Install the radiator stay rod to the radiator and at the cowl. Connect the gage oil line at the flexible connection on the left side of the cylinder block. Connect the choke and throttle cables at the carburetor. Connect the horn wire at the bottom of the steering post. Connect the two stop light wires at the hydraulic brake master cylinder. Connect the blackout light wire at the connection at the cowl. Connect the three headlight wires on the left fender to junction block at the left side of the cowl. Connect the ground strap leading from the rear of the cylinder head to the cowl. Connect the battery cable at the starting motor switch. Connect the wire leading from the coil to the junction block at the right-hand side of the cowl. Connect the wire leading from the junction block at the left side of the cowl to the generator regulator. Connect the positive cable at the battery. Connect the fuel line at the fuel filter.

f. Install Clutch and Brake Pedal Pads and Steering Wheel. Place the clutch and brake pedal pads in the clutch and brake pedals so that the raised ends of the pedal pads are toward the steering post. Install the two cap screws that secure the pedals to the levers. Install the steering wheel on the steering shaft. Connect the foot accelerator to the accelerator rod. Install the transmission gear shift lever on the transmission.

g. Install Air Cleaner. Install the air cleaner with four wing nuts securing it to the brackets on the cowl. Connect the air cleaner hose to the air cleaner.

h. Install Hood and Windshield. Place the hood in position on the vehicle and install the five cap screws that secure the hood to the cowl. Place the windshield in position on the cowl and install the wing nuts that secure the windshield at each side of the cowl.

Section IV

FRAME

43. INSPECTION BEFORE REMOVAL.

a. Position the vehicle on a clean level floor. Attach a plumb bob to the grease fittings at the forward ends of the front spring shackle brackets. Mark the floor at the point indicated by the plumb bob. Attach the plumb bob to the grease fittings at the rear of the rear spring shackle brackets and mark the floor at the point indicated by the plumb bob. Move the vehicle off the markings on the floor. Measuring from the markings on the floor, compare the distance between the front shackle and the rear shackle on the one side of the vehicle with the same measurement on the opposite side of the vehicle, and compare the diagonal distance between each of the front shackles and the rear shackles on the opposite side of the vehicle. Differences of more than ¼ inch in these measurements indicate misalinement that must be corrected. If these comparative measurements are the same within ¼ inch, the frame is not misalined.

44. REMOVAL.

a. Remove Battery (fig. 101). Disconnect the positive and negative cables from the battery. Remove the two wing nuts that secure the battery hold-down rack and remove the rack from the battery.

BODY AND FRAME

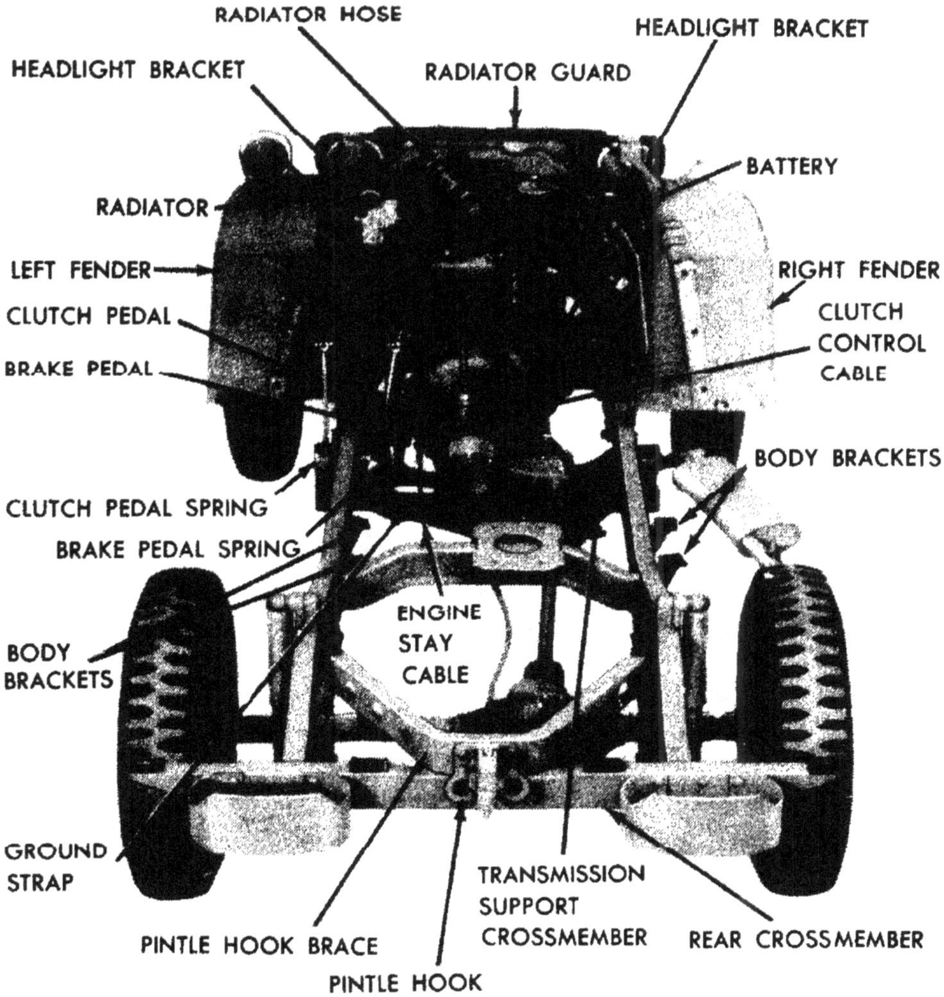

Figure 103 — Chassis

b. **Remove Body and Fenders.** Remove the body (par. 40). Loosen the wing nuts that secure the headlight support bracket to the top of the fenders. Remove the seven bolts that secure the right-hand fender to the frame and to the radiator guard and remove the fender. Remove the eight bolts that secure the left fender to the frame and to the radiator guard, and remove the fender.

c. **Remove Radiator and Radiator Guard** (fig. 103). Drain the coolant from the radiator. Disconnect the hose connections at the top and bottom of the radiator. Remove the two bolts that secure the radiator to the frame front crossmember and remove the radiator. Remove the three nuts that secure the radiator guard to the frame front crossmember and remove the radiator guard.

855473 O - 49 - 9

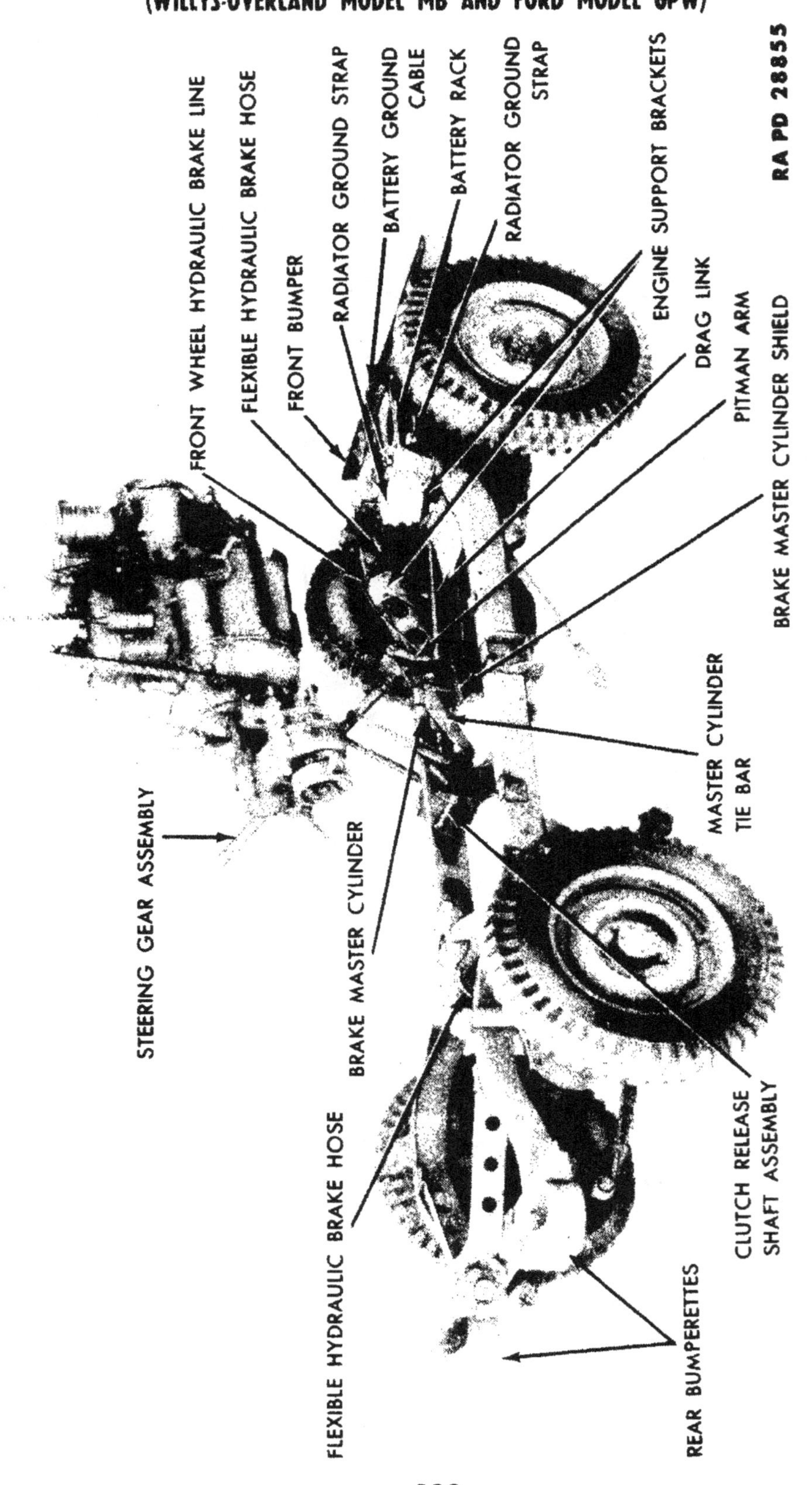

Figure 104 — Removing Engine From Chassis

d. Remove Exhaust Pipe and Transmission Shield (fig. 100). Remove the two bolts that secure the exhaust pipe at the exhaust manifold. Remove the five bolts that secure the transmission shield to the transmission support crossmember. Remove the exhaust pipe and shield from the vehicle.

e. Disconnect the Propeller Shafts (fig. 100). Remove the two U-bolts that secure the front propeller shaft to the transfer case. Remove the propeller shaft from the universal joint flange on the transfer case. Remove the four nuts that secure the rear propeller shaft to the transfer case. Remove the propeller shaft from the transfer case.

f. Remove Front and Rear Bumpers (fig. 104). Remove the eight nuts and bolts that secure the two rear bumperettes to the rear crossmember of the frame and remove the bumperettes. Remove the four nuts and bolts that secure the front bumper bar to the frame and remove the bumper bar.

g. Remove Engine From Vehicle (figs. 103 and 104). Remove the two nuts that secure the transmission to the transmission support crossmember. Remove the ground strap leading from the transmission support crossmember to the transmission. Remove the two nuts that secure the engine stay cable to the transmission support crossmember. Remove the clevis pin that secures the clutch control cable to the clutch release shaft. Remove the two nuts and cap screws that secure the engine mounting to the engine support brackets on each side of the frame. Remove the two ground straps leading from both engine support brackets to the engine. Install a suitable lifting sling on the engine and remove the engine from the frame.

h. Remove Steering Gear Assembly (fig. 104). Disconnect the drag link at the Pitman arm. Remove the cap screw that secures the brake master cylinder shield to the master cylinder. Remove the three nuts and bolts that secure the steering gear assembly to the frame. Remove the steering gear assembly and master cylinder shield from the vehicle.

i. Remove Clutch and Brake Levers (fig. 103). Remove the cotter pin from the clutch and brake pedal shaft at the clutch end of the shaft. Remove the nut and bolt that secure the clutch pedal to the shaft. Remove the clutch pedal and Woodruff key from the shaft. Remove the clutch and brake pedal return springs leading from the pedals to the transmission support crossmember. Remove the two cap screws that secure the master cylinder tie bar to the brake master cylinder. Remove the clutch and brake pedal shaft with the brake pedal and the clutch release shaft from the frame as a unit.

j. Remove Brake Master Cylinder (fig. 104). Disconnect the hydraulic brake line leading to the flexible connection at the forward

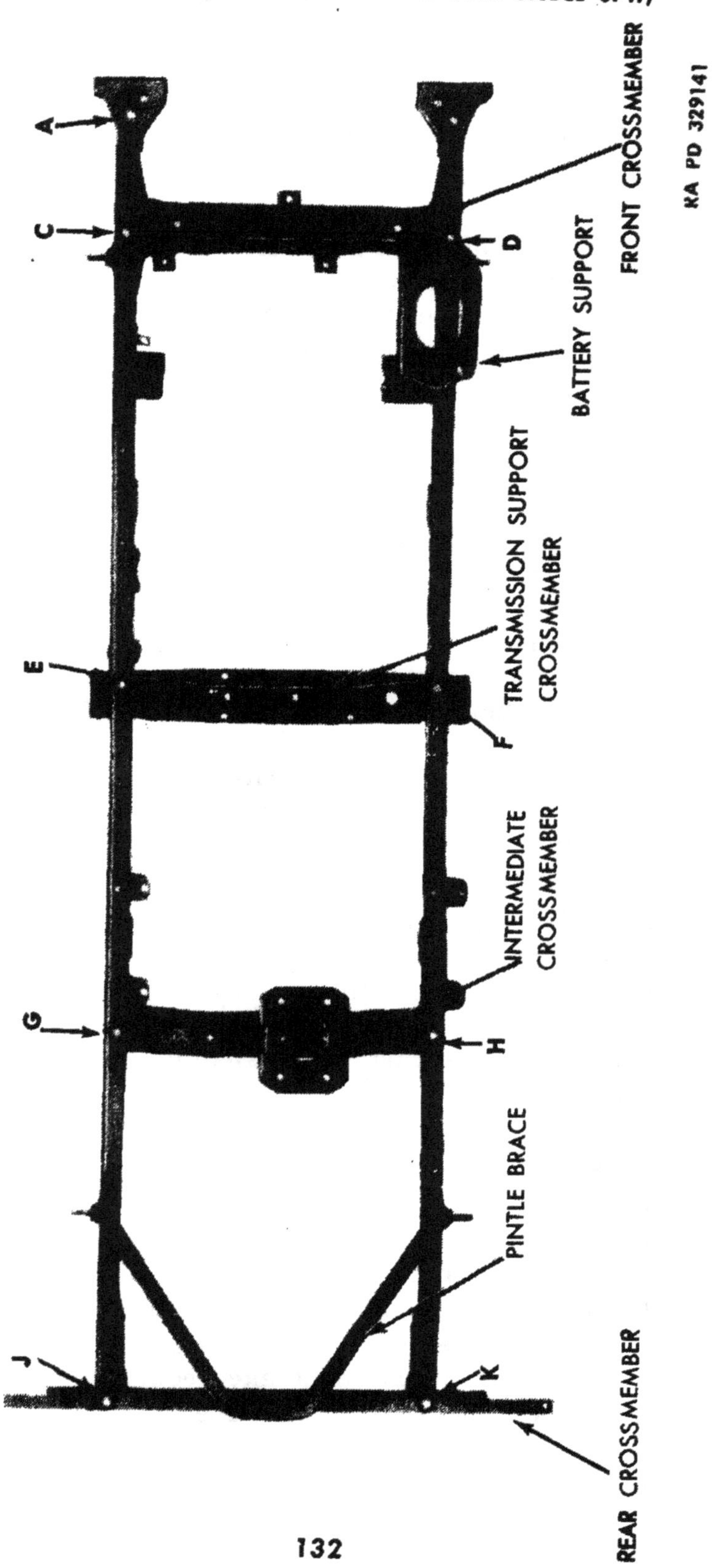

Figure 105 — Frame Assembly

end of the frame. Remove the lock plate that secures the flexible hose to the frame. Disconnect the hydraulic brake line leading to the rear wheels at the master cylinder. Remove the two bolts that secure the master cylinder to the frame and remove the master cylinder. Disconnect the hydraulic brake line at the flexible connection on the center crossmember. Remove the lock plate that secures the flexible hose to the center crossmember.

k. Remove Front Axle (fig. 35). Remove the cotter pin from the upper end of the two front shock absorbers. Pull the shock absorbers off the brackets on the frame. Remove the two spring shackle bushings from the forward end of each front spring and remove the spring shackles. Remove the shackle bolt that secures the rear ends of the front springs to the frame. Remove the front axle from the frame.

l. Remove Rear Axle (fig. 77). Remove the cotter pin from the upper ends of the two rear shock absorbers and pull the shock absorbers off the brackets on the frame. Remove the two spring shackle bushings at the rear end of the frame from each spring. Remove the shackle bolts that secure the forward end of the rear springs to the frame. Remove the rear axle from the frame.

m. Remove Ground Straps (fig. 104). Remove the two radiator ground straps at the front crossmember of the frame. Remove the ground strap at the transmission support crossmember. Remove the battery ground cable at the front crossmember.

45. INSPECTION AND REPAIR.

a. Inspection.

(1) CHECK FRAME ALINEMENT (fig. 105). The extent of misalinement of the frame can be determined by taking measurements at the various points as indicated in figure 105. Measure the distance from A to D and from B to C. The distance between these two points should not vary more than ⅛ inch. Measure the distance from C to F and D to E. The distance between these two points should not vary more than ⅛ inch. Measure the distance from E to H and F to G. The distance between these two points should not vary more than ⅛ inch. Measure the distance from H to J and G to K. The distance between these two points should not vary more than ⅛ inch. If the frame is found to be out of alinement, in most cases it can be corrected by straightening or replacing the damaged frame members or sections.

(2) INSPECT FRAME FOR LOOSE RIVETS. Replace any rivets that are loose or missing. Replace the battery support bracket if bent.

855473 O - 49 - 10

46. INSTALLATION.

a. **Install Ground Straps** (fig. 104). Install the two radiator ground straps at the front crossmember of the frame. Install the ground strap at the transmission support crossmember. Install the battery ground cable at the front crossmember.

b. **Install Rear Axle** (fig. 77). Place the rear axle in position under the frame. Install the rear spring forward shackle bolts with the grease fittings facing outward. Install the two castellated nuts and cotter pins that secure the shackle bolts. Install the rear spring rear shackles with the shackle ends facing outward. Install the two shackle bushings to each spring shackle. Install the shock absorbers to the brackets at each side of the frame.

c. **Install Front Axle** (fig. 35). Place the front axle in position under the frame. Install the two front spring rear shackle bolts with the grease fittings facing outward. Install the two castellated nuts and cotter pins that secure the shackle bolts. Install the spring shackle at the forward end of the frame with the shackle ends facing outward. Install the two spring shackle bushings to each spring shackle. Install the shock absorbers to the brackets at each side of the frame.

d. **Install Brake Master Cylinder** (fig. 104). Install the lock plate that secures the hydraulic brake flexible hose to the center crossmember of the frame. Connect the hydraulic brake line at the flexible connection on the center crossmember. Install the two bolts that secure the hydraulic brake master cylinder to the side of the frame. Install the brake line leading from the rear wheels to the brake master cylinder. Install the lock plate that secures the brake flexible hose at the bracket on the frame. Install the brake line leading from the flexible hose to the master cylinder.

e. **Install Clutch and Brake Pedals** (fig. 103). Install the clutch and brake shaft with brake pedal and clutch release shaft onto the frame. Install the two cap screws that secure the master cylinder tie bar to the master cylinder. Insert the brake pedal rod in the master cylinder. Install the clutch and brake return springs leading from the clutch and brake pedals to the transmission support crossmember. Install the Woodruff key on the clutch and brake pedal shaft. Slide the clutch pedal on the clutch shaft. Install the lock bolt that secures the clutch pedal to the shaft and install the cotter pin.

f. **Install Steering Assembly and Bleed Brakes** (fig. 104). Place the steering assembly in position on the frame. Install the three nuts and bolts that secure the steering assembly to the frame. Connect the drag link at the Pitman arm. Bleed the hydraulic brake system. (Refer to TM 9-803.)

g. **Install Engine in Vehicle** (fig. 104). Install a suitable lifting sling on the engine and place the engine in position on the frame. Move the rear of the engine slightly to the right-hand side so as to allow the ball joint on the transfer case to enter in the clutch release shaft. Install the two nuts and bolts that secure the engine mounts to the engine support brackets on each side of the frame. Install the two ground straps leading from the engine support brackets to the engine. Install the two nuts that secure the transmission to the transmission support crossmember. Install the ground strap leading from the transmission to the transmission support crossmember. Install the clevis pin that secures the clutch control cable to the clutch release shaft. Install the engine stay cable to the transmission support crossmember.

h. **Install Front and Rear Bumpers** (fig. 104). Place the front bumper in position on the frame. Install the four nuts and bolts that secure the bumper to the ends of the frame. Place the two bumperettes in position on the rear crossmember of the frame and install the eight cap screws that secure the bumperettes to the rear crossmember of the frame.

i. **Install Propeller Shafts** (fig. 100). Place the front propeller shaft in the universal joint flange at the transfer case. Install the two U-bolts that secure the propeller shaft to the transfer case. Place the rear propeller shaft yoke flange in position on the transfer case. Install the four nuts that secure the propeller shaft to the transfer case.

j. **Install Exhaust Pipe and Transmission Shield** (fig. 100). Install the two nuts that secure the exhaust pipe to the exhaust manifold. Place the transmission shield in position under the vehicle. Install the five bolts that secure the transmission shield to the transmission. Install the clamp that secures the exhaust pipe to the shield.

k. **Install Radiator and Radiator Guard** (fig. 103). Place the insulators on the radiator brackets on the frame. Place the radiator in position in the brackets on the frame. Install the flat washers, ground straps, and flat washers on the two radiator studs. Install the nuts that secure the radiator to the frame. Install the upper and lower radiator hose connections. Install the radiator guard in the bracket in front of the radiator. Install the three nuts and bolts that secure the guard to the frame.

l. **Install Fenders and Body** (fig. 102). Place the fenders in position on the vehicle. Install the seven cap screws that secure each fender to the radiator guard and to the frame. Install the body on the chassis (par. 41).

m. **Install Battery** (fig. 101). Place the battery in the battery rack. Install the hold-down bracket on the battery. Install the two wing nuts that secure the hold-down bracket. Connect the negative and positive cables to the battery.

n. **Lubricate.** Lubricate the chassis of the vehicle with specified lubricants. Fill the radiator to proper level with coolant.

Section V

FITS AND TOLERANCES

47. FITS AND TOLERANCES.

Fit Location and Name	Manufacturer's Fit Tolerance	Fit Wear Limit	Type of Fit
a. **Front Springs.**			
Spring bushing and shackle bolt	—	0.010 in.	Running
Torque reaction spring and shackle bolt	—	0.010 in.	Running
Torque reaction spring shackle inner bushing and lower shackle bolt	—	0.010 in.	Running
Torque reaction spring shackle outer bushing and lower shackle bolt	—	0.010 in.	Running
b. **Rear Springs.**			
Rear spring bushing and shackle bolt	—	0.010 in.	Running
c. **Steering Gear.**			
Sector shaft bushing and sector shaft	—	0.001 in.	Running
d. **Monroe Shock Absorber.**			
Piston and pressure tube	0.002 in.	0.004 in.	Running

CHAPTER 4

SPECIAL TOOLS

48. PURPOSE.

a. The special tools required for maintenance and repair of the ¼-ton 4 x 4 Truck are listed in SNL G-27.

b. The following list, extracted from SNL G-27, contains those special tools required to perform the operations described in this manual. The list is supplied for identification purposes only; it is not to be used as a basis for requisition.

49. LIST OF SPECIAL TOOLS.

a. Special Tools for Power Train.

Name	Federal Stock Number	Mfr's Number	Figure Number
Gage, drive pinion setting (set)	41-G-176	KM-J-589-S	67 and 106
Locator, idle gear thrust washer	41-L-1570	KM-J-1758	106
Remover, drive pinion flange and side bearing	41-R-2378-30	KM-J-872-S	49 and 106
Remover, drive pinion oil seal	41-R-2378-40	KM-J-1742	106
Remover, front axle outer oil retainer	41-R-2384-38	KM-J-943	55-80
Remover, mainshaft front cone, transmission	41-R-2368-200	KM-J-1749	106
Replacer, axle shaft, front and rear oil seal	41-R-2391-20	KM-J-1753	56 and 106
Replacer, differential side bearing cone	41-R-2391-65	KM-J-1763	68 and 106
Tool, oil seal assembly shifter shaft (set)	41-T-3280	KM-J-1757	106
Tool, universal joint, assembly and disassembly	41-T-3379	KM-J-881-A	—
Wrench, wheel bearing nut	41-W-3825-200	GP-17033	—

b. Special Tools for Shock Absorbers (Monroe).

Name	Federal Stock Number	Mfr's Number	Figure Number
Compressor, shock absorber grommet	41-C-2554-400	MAS-1148	—
Filler cup, Monroe shock absorber	41-F-2985-200	—	106
Remover, base valve	41-R-2373-340	—	106
Remover, rod guide and seal assembly	41-R-2373-115	—	106
Replacer, pressure tube	41-R-2399-350	—	106
Spanner wrench, special piston rod guide and seal assembly	41-W-3336-745	—	88 and 106
Thimble, piston rod	41-T-1657	—	106

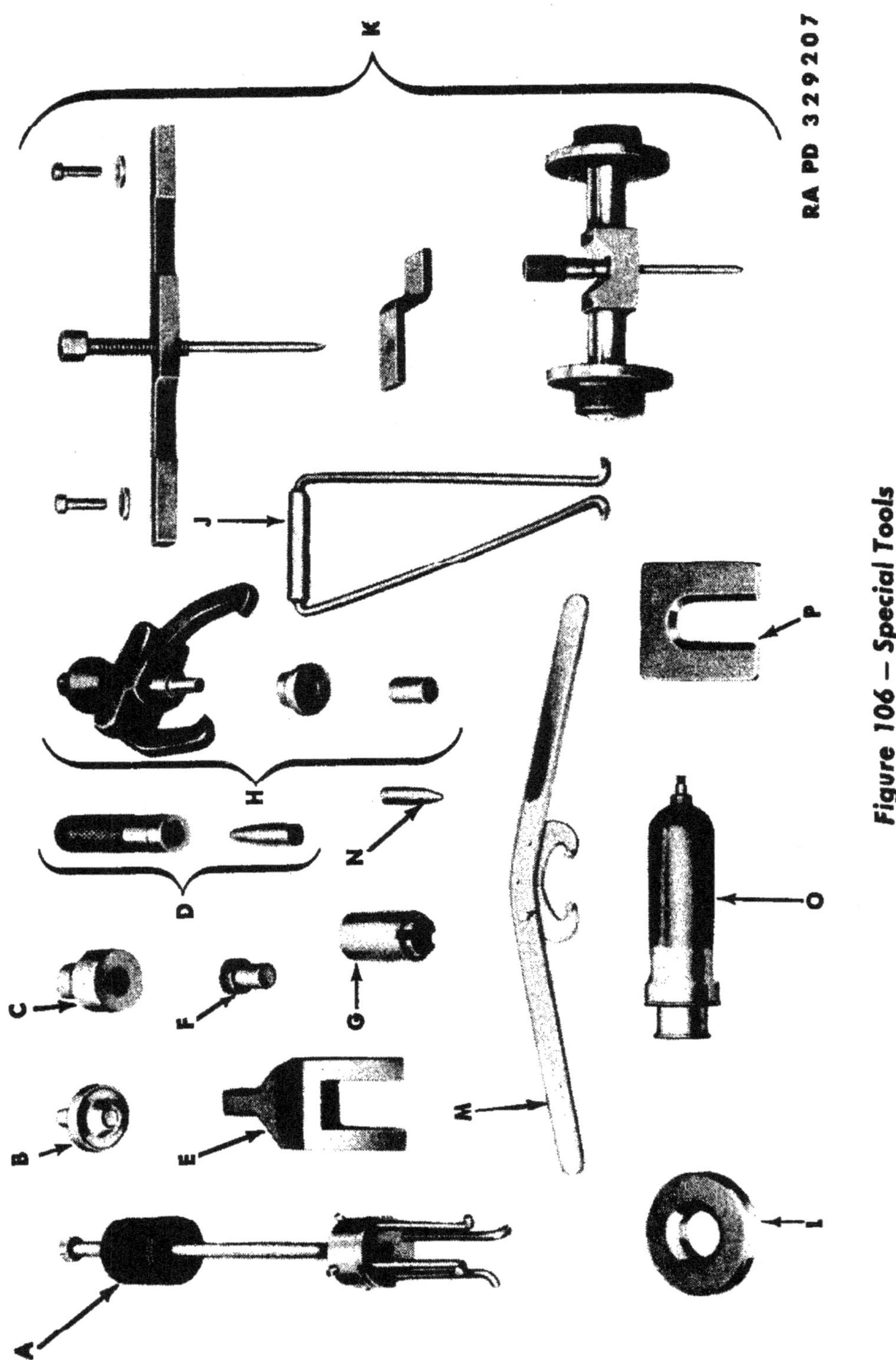

Figure 106 — Special Tools

	Federal Stock No.
A—REMOVER, DRIVE PINION OIL SEAL	41-R-2378-40
B—REPLACER, AXLE SHAFT FRONT AND REAR OIL SEAL	41-R-2391-20
C—REPLACER, DIFFERENTIAL SIDE BEARING CONE	41-R-2391-65
D—TOOL, OIL SEAL ASSEMBLY, TRANSFER CASE SHIFTER SHAFT SET	41-T-3280
E—REMOVER, MAINSHAFT FRONT CONE	41-R-2368-200
F—LOCATOR, IDLER GEAR THRUST WASHER	41-L-1570
G—REPLACER, PRESSURE TUBE	41-R-2399-350
H—REMOVER, DRIVE PINION FLANGE AND DIFFERENTIAL SIDE BEARING	41-R-2378-30
J—REMOVER, FRONT AXLE OUTER OIL RETAINER	41-R-2384-38
K—GAGE, DRIVE PINION SETTING (SET)	41-G-176
L—REMOVER, ROD GUIDE AND SEAL ASSEMBLY	41-W-2373-115
M—SPANNER WRENCH, SPECIAL PISTON ROD GUIDE AND SEAL ASSEMBLY	41-W-3336-745
N—THIMBLE, PISTON ROD	41-T-1657
O—FILLER CUP, SHOCK ABSORBER	41-F-2985-200
P—REMOVER, PRESSURE VALVE	41-R-2373-340

RA PD 329207-B

Legend for Figure 106 — Special Tools

ORDNANCE MAINTENANCE — POWER TRAIN, BODY, AND FRAME FOR ¼-TON 4 x 4 TRUCK (WILLYS-OVERLAND MODEL MB AND FORD MODEL GPW)

REFERENCES

PUBLICATIONS INDEXES.

The following publications indexes should be consulted frequently for latest changes or revisions of references given in this section and for new publications relating to materiel covered in this manual:

a. Introduction to Ordnance Catalog (explaining SNL system) .... ASF Cat. ORD 1 IOC

b. Ordnance Publications for Supply Index (index to SNL's) .... ASF Cat. ORD 2 OPSI

c. Index to Ordnance Publications (listing FM's, TM's, TC's, and TB's of interest to ordnance personnel, OPSR, MWO's, BSD, S of SR's, OSSC's, and OFSB's; and includes Alphabetical List of Major Items with Publications Pertaining Thereto) .... OFSB 1-1

d. List of Publications for Training (listing MR's, MTP's, T/BA's, T/A's, FM's, TM's, and TR's concerning training) .... FM 21-6

e. List of Training Films, Film Strips, and Film Bulletins (listing TF's, FS's, and FB's by serial number and subject) .... FM 21-7

f. Military Training Aids (listing Graphic Training Aids, Models, Devices, and Displays) .... FM 21-8

STANDARD NOMENCLATURE LISTS.

Cleaning, preserving and lubricating materials; recoil fluids; special oils, and miscellaneous related items .... SNL K-1

Ordnance maintenance sets .... SNL N-21

Soldering, brazing and welding materials, gases and related items .... SNL K-2

Tools, maintenance for repair of automotive vehicles SNL G-27 Volume 1

Tool-sets, for ordnance service command automotive shops .... SNL N-30

Tool-sets, motor transport .... SNL N-19

Truck, ¼-ton, 4 x 4, command reconnaissance (Ford and Willys) .... SNL G-503

REFERENCES

EXPLANATORY PUBLICATIONS.

Fundamental Principles.

Automotive brakes	TM 10-565
Automotive electricity	TM 10-580
Automotive power transmission units	TM 10-585
Basic maintenance manual	TM 38-250
Chassis, body, and trailer units	TM 10-560
Electrical fundamentals	TM 1-455
Military motor vehicles	AR 850-15
Motor vehicle inspections and preventive maintenance service	TM 9-2810
Precautions in handling gasoline	AR 850-20
Sheet metal work, body, fender, and radiator repairs	TM 10-450
Standard military motor vehicles	TM 9-2800
The body finisher, woodworker, upholsterer, painter, and glassworker	TM 10-455
The machinist	TM 10-445

Maintenance and Repair.

Cleaning, preserving, lubricating and welding materials and similar items issued by the Ordnance Department	TM 9-850
Cold weather lubrication and service of combat vehicles and automotive materiel	OFSB 6-11
Maintenance and care of pneumatic tires and rubber treads	TM 31-200
Ordnance Maintenance: Electric equipment (Auto-Lite)	TM 9-1825B
Ordnance Maintenance: Engine and engine accessories for ¼-ton 4 x 4 truck (Ford and Willys)	TM 9-1803A
Ordnance Maintenance: Hydraulic brake system (Wagner)	TM 9-1827C
Ordnance Maintenance: Speedometers and tachometers (Stewart-Warner)	TM 9-1829A

Operator's Manual.

¼-ton 4 x 4 truck (Willys-Overland model MB and Ford model GPW)	TM 9-803

Protection of Materiel.

Camouflage .... FM 5-20

Chemical decontamination, materials and equipment .... TM 3-220

Decontamination of armored force vehicles .... FM 17-59

Defense against chemical attack .... FM 21-40

Explosives and demolitions .... FM 5-25

Storage and Shipment.

Ordnance storage and shipment chart, group G—Major items .... OSSC-G

Registration of motor vehicles .... AR 850-10

Rules governing the loading of mechanized and motorized army equipment, also major caliber guns, for the United States Army and Navy, on open top equipment published by Operations and Maintenance Department of Association of American Railroads.

Storage of motor vehicle equipment .... AR 850-18

INDEX

ORDNANCE MAINTENANCE — POWER TRAIN, BODY, AND FRAME FOR ¼-TON 4 x 4 TRUCK (WILLYS-OVERLAND MODEL MB AND FORD MODEL GPW)

B — Cont'd

C

D

INDEX

ORDNANCE MAINTENANCE — POWER TRAIN, BODY, AND FRAME FOR ¼-TON 4 x 4 TRUCK (WILLYS-OVERLAND MODEL MB AND FORD MODEL GPW)

INDEX

S — Cont'd

T

U

W

Y

U. S. GOVERNMENT PRINTING OFFICE : O—1949

www.ingramcontent.com/pod-product-compliance
Lightning Source LLC
Chambersburg PA
CBHW080417030426
42335CB00020B/2490